T0255983

Forest Family

Critical Plant Studies

PHILOSOPHY, LITERATURE, CULTURE

Series Editor

Michael Marder
(*IKERBASQUE*/*The University of the Basque Country, Vitoria*)

VOLUME 4

The titles published in this series are listed at *brill.com/cpst*

Forest Family

Australian Culture, Art, and Trees

Edited by

John C. Ryan and Rod Giblett

BRILL

RODOPI

LEIDEN | BOSTON

Cover illustration: Marianne North. Karri Gums near the Warren River, West Australia. Painting 782. Reprinted with Permission from Royal Botanic Gardens, Kew.

Library of Congress Cataloging-in-Publication Data

Names: Ryan, John (John Charles) (Poet), editor.
Title: Forest family : Australian culture, art, and trees / edited by John C.
    Ryan and Rod Giblett.
Description: Leiden ; Boston : Brill Rodopi, [2018] | Series: Critical plant
    studies: philosophy, literature, culture, ISSN 2213-0659 ; volume 4 |
    Includes bibliographical references.
Identifiers: LCCN 2018019416 (print) | LCCN 2018022010 (ebook) | ISBN
    9789004368651 (E-book) | ISBN 9789004368644 (hardback : alk. paper)
Subjects: LCSH: Old growth forests—Western Australia—History. | Human-plant
    relationships—Western Australia—History. | Forests in literature.
Classification: LCC SD387.O43 (ebook) | LCC SD387.O43 F67 2018 (print) | DDC
    333.7509941—dc23
LC record available at https://lccn.loc.gov/2018019416

Typeface for the Latin, Greek, and Cyrillic scripts: "Brill". See and download: brill.com/brill-typeface.

ISSN 2213-0659
ISBN 978-90-04-36864-4 (paperback)
ISBN 978-90-04-36865-1 (e-book)

To the Forest and the Family

∴

Contents

Acknowledgements

Our sincere thanks go to Michael Marder, editor of Brill's "Critical Plant Studies" series, for his encouragement at the outset and to Meghan Connolly of the same press for her support of the project. As editors, we extend gratitude to our contributors, Nandi Chinna, Robin Ryan, and Juha Tolonen, for making *Forest Family* a truly distinctive and multimodal contribution to studies of Australian forests within the exciting framework of critical plant studies.

John Charles Ryan acknowledges the Department of English and Cultural Studies at the University of Western Australia, Perth, for his appointment as Honorary Research Fellow (2017–2020) and the School of Arts at the University of New England, Armidale, Australia, for a Postdoctoral Research Fellowship (2017–2020). He also thanks the staff members of the Northcliffe Visitor Center and Fiona Sinclair of Southern Forest Arts for generously suggesting artists and artworks to include in Chapter 6. A condensed version of Chapter 6 was published as "'Vegetable Giants of the West': Plant Ethics in the Photography of Australian Karri (*Eucalyptus diversicolor*) Forests, 1890 to the Present" in the *Journal of the European Association for Studies of Australia* (2016, 7.1: 61–83). John thanks the editor of the journal, Martina Horáková, and two anonymous referees for critical feedback that enhanced the discussion.

Rod Giblett is grateful to the Center for Research in Entertainment, Arts, Technology, Education, and Communications (CREATEC) at Edith Cowan University for funding several research assistants, initially Christine Barton and John C. Ryan, and later Nandi Chinna and Juha Tolonen. An earlier version of Chapter 4 was first published as "Family Trees: Jarrah, Karri and the Gibletts of Manjimup" in *Traces of an Active and Contemplative Life: 1983–2013* (Champaign, IL: Common Ground, 2013, 175–187). He is also grateful to Gail Ipsen Cutts, Director of Community Services, Shire of Manjimup, for her advice and suggestions.

List of Contributors

Nandi Chinna

is a research consultant, poet, and activist. Her poetry publications include *Swamp*: *Walking the Wetlands of the Swan Coastal Plain* (Fremantle Press, 2014), *Alluvium* (with illustrator Andrea Smith, Lethologica Press, 2012), *How to Measure Land* (Picaro Press, Byron Bay Writers Festival Poetry Prize winner, 2010), and *Our Only Guide is Our Homesickness* (Five Islands Press, 2007). Her latest poetry collection, *The Future Keepers*, is forthcoming from Fremantle Press in 2019. In 2016 Nandi was Writer in Residence at Kings Park and Botanic Gardens in Perth, Western Australia. She won the 2016 Fremantle History Award for her history of Clontarf Hill, and was shortlisted for the 2016 Red Room Poetry Fellowship. For the 2017 Perth International Arts Festival, she collaborated with Amy Sharrocks and the Museum of Water on a community water poem and an interactive walking tour of lost water bodies. She facilitates regular writing labs on "poepatetics"—the art of walking and writing.

Rod Giblett

is a great-great grandson of John and Ann Giblett, the first European settlers of Balbarrup and Dingup in the Manjimup area of Western Australia. For a long time, he has had an interest in his family history, in the environmental history and aesthetics of the forests of Southwest Australia, and in the cultural and natural history of jarrah and karri trees, the two most visibly dominant tree species of the two bioregions to which they give their names. He has written books about people and place, as well as books of nature writing, ecocriticism, and environmental philosophy, history, and theology, including *Environmental Humanities and Theologies: Ecoculture, Literature and the Bible* (Routledge, 2018). He also lived by a wetland for many years and wrote and published many books about it and other wetlands. He is Honorary Associate Professor of Environmental Humanities at Deakin University.

John Charles Ryan

is a poet and scholar who holds appointments as Postdoctoral Research Fellow in the School of Arts at the University of New England in Australia and Honorary Research Fellow in the School of Humanities at the University of Western Australia. His teaching and research cross between the environmental and digital humanities. He is the author or editor of several research books, including *Digital Arts: An Introduction to New Media* (Bloomsbury, 2014, as co-author), *The Language of Plants: Science, Philosophy, Literature* (University of

Minnesota Press, 2017, as co-editor and contributor), *Plants in Contemporary Poetry: Ecocriticism and the Botanical Imagination* (Routledge, 2017, as author), and *Southeast Asian Ecocriticism: Theories, Practices, Prospects* (Lexington Books, 2017, as editor and contributor).

   Robin Ryan
studied music at the University of Western Australia and the University of Washington, Seattle, USA. She contributed the first Master's thesis on urban Aboriginal music through Monash University, where her research into the leaf music of Indigenous Australians (PhD, 1999) led to employment as a specialist adviser to *Currency Companion to Music and Dance in Australia* (2003). A former Research Fellow at Macquarie University, Sydney (2001–2005), Robin is currently affiliated with the Western Australian Academy of Performing Arts (WAAPA) at Edith Cowan University. Some of her recent research is published in *Current Directions in Ecomusicology: Music, Culture, Nature* (Routledge) and *Collaborative Ethnomusicology* (Lyrebird Press); and the journals *Environmental Humanities, Perfect Beat, M/C Journal, Societies, Sound Scripts*, and the *Journal of Music Research Online*. Robin currently resides in the south-east Australian forest port of Eden, New South Wales.

   Juha Tolonen
is an artist, academic, and entrepreneur. His work can be found in the public collections of the Australian Parliament and Art Gallery of Western Australia. He is an Adjunct Lecturer at Edith Cowan University and is the co-author of *Photography and Landscape* published by Intellect Books. Juha currently lives and works in Lapland innovating new products in nature-based tourism.

CHAPTER 1

Introducing *Forest Family*

John C. Ryan and Rod Giblett

Forest Family arose initially out of the interest of Rod Giblett in the early pioneering history of his family during the mid-nineteenth century in the southwest forests of Western Australia. The book also arose out of a desire not to write the typical kind of family history that would only appeal to other members of the family. In general, family histories focus exclusively on people, and not on the places and their plants and animals that shaped and affected the family and its history. Such histories tend to ignore or downplay the plants, animals, and places that are agents and players in the family history. These might only have supporting or walk-on roles in the story, and the natural environment might only provide a backdrop against which human action takes place.

As both editors have a longstanding interest in, and commitment to, the interactions between people and places, and between them and plants and animals, in natural and environmental history, in the relationship and interactions between nature and culture, in nature conservation and cultural preservation, and in critical plant studies and the transdisciplinary environmental humanities, *Forest Family* is—and was always going to be—a different sort of book than your run-of-the-mill family history.

Family history as related in chapters 4 and 5 of *Forest Family* is thus interwoven with natural history of the jarrah and karri ecosystems that are essential to the narrative. As presented in these chapters, human history is part and parcel of the environmental history of the Southwest Australian botanical province. This is the biome where the jarrah and karri forests segue from one to the other, from one bioregion and river catchment to the next. Both editors have a longstanding interest in, and commitment to, the valuing of bioregional catchments as the ambit of the home habitat extending the family home to the transhuman community.

As members of the Giblett family were early pioneers and builders in the nineteenth century of their own abodes and civic structures, such as a house, garden, church, and bridge, *Forest Family* weaves together built and landscape architectural and engineering history in the account of the heritage and cultural history of built structures in chapter 5. These are also depicted pictorially in the photo essay by Juha Tolonen.

© KONINKLIJKE BRILL NV, LEIDEN, 2018 | DOI 10.1163/9789004368651_002

The magnificent jarrah and karri trees and forests were also the subject of, and inspiration for, artistic and creative expression in painting, poetry, fiction, film, botanical illustration and music in the nineteenth and twentieth centuries. Art and literary history interleave with environmental and natural history in chapters 6 and 7 of *Forest Family*, and all of them with family history in chapters 4 and 5. Art and literary history become an essential part of environmental and natural history in *Forest Family* rather than remaining separate.

As a forest block in this region named after the pioneering Giblett family was the site of a controversial campaign in the 1990s to stop logging, chapter 8 also discusses conservation history in relation to environmental and natural history. This chapter makes use of oral history to understand these broader contexts, as many of the leading activists in the blockade were interviewed for this chapter and tell their personal stories in it.

Weaving together local, oral, natural and environmental history in *Forest Family* follows in the footsteps of similar books, such as Henry David Thoreau's *Walden*, Rod Giblett's *Forrestdale: People and Place* (2006) and John Ryan's *Green Sense: The Aesthetics of Plants, Place, and Language* (2012). Both *Green Sense* and *Forrestdale* are based on oral history interviews and Thoreau related conversations with his friends and neighbors. *Walden* is much more autobiographical and philosophical than *Forest Family* though. *Walden* is also much more concerned than *Forest Family* with the details of everyday life and with observations of nature based on a journal that Thoreau kept for many years before and after his sojourn by Walden Pond.

Both editors of *Forest Family* have not lived in the region discussed in it, unlike Thoreau who lived for a couple of years by Walden Pond and lived for most of his life in the region. Like Thoreau, Rod Giblett lived for a long time in one place by a lake and kept a nature journal for a couple of years that morphed into the book *Black Swan Lake: Life of a Wetland* (2013). Yet *Walden*, like *Forest Family*, weaves together family history, natural history of species, environmental history of Walden Pond and Woods, architectural history of the one-room house he built, and conservation history of the clearing of trees and his opposition to it and calls for environmental protection.

Forest Family responds, furthermore, to the status of plant life in an increasingly globalized world where there is less room for ancient forests to thrive. In its analysis of thirteen-thousand botanical taxa worldwide, for example, the International Union for Conservation of Nature calculated a sixty-eight percent rate of critical endangerment. *The State of the World's Plants Report* issued by Kew Gardens suggests that twenty-five percent of the world's plants are threatened with extinction. When we consider the profound interconnections between flora and humanity through food, drink, medicine, fiber, elemental exchange, cultural heritage, and spiritual sustenance, these statistics become especially harrowing.

Human modification of old-growth ecosystems clearly triggers effects at local and global scales. Old-growth is known for harboring biological diversity, sequestering carbon, regulating hydrological processes, and providing a wealth of other ecosystemic functions. Large intact forests often straddle boundaries of different kinds, highlighting the need for efforts that are grounded in, but extend beyond, bioregions (Potapov et al. 2017). Yet, frequently missing from enumerations of the value of old-growth ecosystems are the artistic, cultural, literary, and political agencies of ancient trees. This includes how primordial trees inspire the human imagination while mediating diverse aspects of human and non-human existence.

Book Structure and Chapter Overviews

The contributors to this volume develop an outlook on old-growth as a collective of beings, as a transhuman, multispecies "family" consisting of human and non-human members. In broad terms, *Forest Family* argues that human history, culture, art, and politics are integrally related to old-growth habitats. The human, other-than-human animal, and the forest shape each other dialogically and reciprocally. In this regard, the contributors interpret *family* not as a metaphor deployed to lyricize primordial karri and jarrah forests but as an actual, material, multi-, inter- and cross-species dynamic that has molded the Southwest Australian landscape into what it is today. Along with its Indigenous resonances—outlined with respect to Nyoongar culture in chapter 3 and elsewhere—the notion also recurs in recent writings about Australia by authors with Anglo-European backgrounds.

A notable case in point is Tim Winton's essay "The Island Seen and Felt: Some Thoughts about Landscape," originally published in his book *Island Home* (2015). Winton begins with the qualification that "the island continent has not been mere background. Landscape has exerted a kind of force upon me that is every bit as geological as family. Like many Australians, I feel this tectonic grind—call it a familial ache—most keenly when abroad" (Winton 2017). Family is more than metaphorical. The experience of family is mediated by the body as "this country leans in on you. It weighs down hard. Like family. To my way of thinking, it is family" (Winton 2017). The sense of intimate fellowship with the natural world is the basis for "why we write about it. This is why we paint it. From love and wonder, irritation and fear, hope and despair; because, like family, it refuses to be incidental" (Winton 2017).

Winton's statements are compatible with the core premise of *Forest Family*. The book features seven chapters and a photographic essay by five contributors focusing on the deep cultural resonances of karri and jarrah trees and forests. The volume attempts to bring together the natural and cultural histories of old-growth through a range of content from precolonial to contemporary periods. Discussion of paintings, photography, sculpture, music, performance, historical documents, and literary works depicting Southwest forests provides the foundation of the book's critical approach to old-growth and the discourses surrounding it. *Family Forest* centers on the relationship between vegetal nature and human culture through the narrative vehicle of the Giblett family of Manjimup, Western Australia, and the tall eucalypt forests in which they settled (and which they settled).

The volume embraces the different and interacting histories of the Southwest region and its unique forests—of the past and present—and celebrates this integrated human-plant story for the future. Chapter 2, "From Understory to Overstory: Critical Studies of Old-Growth Trees and Forests" proposes *critical old-growth studies* as an approach to the aesthetic, cultural, economic, spiritual, and symbolic meanings attributed to ancient forest ecosystems and the trees contained within them. Marshalling a wide range of material in the chapter, John Ryan argues that—from the sacred to the ecological, the aesthetic to the pecuniary, the personal to the political—writers, artists, and other commentators over time have highlighted the heterogeneous and, at times, irreconcilable values of old-growth habitats.

At best, old-growth is a contested concept, characterized by multiple perspectives—sometimes overlapping, sometimes repelling—on the vegetal world. An understanding of old-growth as sacred topos and intelligent being-in-the-world furnishes the foundation for a critical approach to the representations, discourses, and aesthetics of primordial forests. Chapter 3, "Forest Giants: Locating Southwest Australian Old-Growth Country," further develops these ideas in relation to the eco- and biocultural milieu of the volume: the Southwest Botanical Province of Western Australia and, more specifically, the Warren Bioregion of which karri country is part.

In chapter 4, "Family Trees: Jarrah, Karri, and the Gibletts of the Balbarrup-Dingup Area," Rod Giblett uses family history as a storied thread linking forest and family together. The chapter draws on a range of archival and contemporary sources, such as the cultural and natural history of jarrah and karri trees and forest, Western Australian botanical texts, and forest management policy. Giblett traces aspects of the biocultural history of the lower Southwest region, centering on the hamlets of Balbarrup and Dingup near the town of Manjimup, and on the Giblett family who were amongst the first European settlers of this area. Giblett family history is related briefly, and is then placed within the context of the natural and environmental history of the eucalypt jarrah and karri environments of the area. Chapter 4 argues for a dialogics of vegetal nature and human culture within which the natural environment and non-human context shaped family history and which the family shaped by felling trees, clearing bush, building houses and churches, farming land, and so on.

In chapter 5, "Built in the Forest: A Hamlet History of Giblett Cultural Heritage," Rod Giblett describes more fully the early Western Australian Gibletts who were the first European settlers and builders in the local Manjimup area; as well as the Muirs who were the first European settlers further east in the larger Warren district. The Gibletts were prolific builders who constructed Dingup House and Dingup Church; Balbarrup House and Balbarrup Post Office; flour mills on Wilgarup River and Channybearup Brook; and the Stone House at Quanuup near Lake Jasper. All except for the last building, these buildings made up the hamlets of Balbarrup and Dingup. The Gibletts also built structures further afield, such as One Tree Bridge over the Donnelly River west of Manjimup. Dingup House and Church remain standing and functioning to this day and are heritage-listed.

Dingup House dates from around 1870 and now operates as a Bed and Breakfast. Dingup Church was built in 1895 and is well-maintained and open to the public. Balbarrup Post Office was opened in 1864 and is commemorated by a cairn with a plaque. Some remnant sections of the original One Tree Bridge are displayed near the current traffic bridge over Donnelly River. The Giblett builders are buried in the old, pioneer Balbarrup cemetery on Perup Road over the other side of Balbarrup Brook from the Post Office. In this chapter Giblett, a direct descendant of the pioneers, examines the history of this cultural heritage and places it in the context of the natural heritage of Southwest forests. The chapter argues that these buildings, their gardens, and the hamlet which they made up were a bit of England transported to the forest in order to both carve out a home in it and to protect themselves from its unhomeliness.

In chapter 6, "From Burls to Blockades: Artistic Interpretations of Karri Trees and Forests," John Ryan analyzes artistic representations of karri trees and forests through the colonial era (1829–1901), early twentieth century (1901–45), late twentieth century (1945–88), and contemporary period (1988–present). Karri trees and forests have captivated photographers, painters, and other artists since the colonial beginnings of Western Australia. The charisma of *Eucalyptus diversicolor*—its remarkable size, striking verticality, trunk textures, color patterns—continues to inspire and challenge today's artists attempting to devise vocabularies for translating their experiences of karris to a creative medium. Whereas historical commentators have been inclined to dismiss the aesthetic virtues of jarrahs and marris, with their wild asymmetries and strange exudations, karris have been extolled in more consistent terms for having classically beautiful qualities: smoothness, sleekness, gracefulness, grandeur, sublimity.

As one of the tallest eucalypt species in the world, second only to Victoria and Tasmania's mountain ash (*Eucalyptus regnans*), the karri tree—as evident in its earliest written and visual representations—enlivens a public environmental imagination that longs for solitude, serenity, and a glimpse of the divine in nature. In stark contrast to the appreciation of karris as inspirational beings, however, late nineteenth- and early-twentieth-century photographs of surveying and logging activities convey a much different story. The perspective on karris, instead, involves instrumental desires where massive old-growth trees are resources to be exploited or behemoths to be overcome for the sake of settler progress.

Chapter 7, "Sing the Karri, Sculpt the Jarrah: Sustaining Old-Growth Forest through the Arts" by musicologist Robin Ryan presents the arts as a medium for ecological relationality and multispecies community through a focus on the Understory Art in Nature Trail adjoining the small town of Northcliffe, situated forty-four kilometers geographically south of Manjimup. For over a decade now, this majestic Southwest forest setting has nurtured the creativity of artists, writers, and musicians working symbiotically with a pristine landscape and dialogically with Indigenous and settler-descendant senses of place. The works are inextricably linked to the history of Northcliffe's ill-conceived colonialist foundation, its heated turn-of-the-century logging dispute and consequent industrial decline, and—more recently—its close escape from a colossal bushfire.

Robin Ryan considers art's contribution to the respectful conservation of land through intentionally cultivated inspiration and minimalist physical intervention. Based on the premise that the iconic trees karri and jarrah possess their own artistic agency, tall timber country imagining in the album *Canopy: Songs for the Southern Forests*, released by Southern Forest Arts in 2006, is critiqued in counterpoint to selected works featured along the meandering 1.2 kilometer sculpture trail. Adding to its economic benefit, the culturalized narrative of nature contributes laterally—through ways that only artworks can—to sustaining old-growth forest.

Finally, chapter 8 "Old-Growth Activism: The Giblett Forest Rescue of 1994 and 1997" by Nandi Chinna provides the first comprehensive account of the seminal forest protection campaigns of the 1990s. The Giblett forest blockade was a momentous event in the history of conservation in Western Australia as it was one of the first protests to successfully protect an old-growth forest slated for logging. During the 1990s, the Giblett Block became the site of one of the state's first long-standing forest protest blockades, an action which was to change public perceptions of forests and forestry practices, instigate previously unheard of public participation and consultation in forestry management issues, and, to some degree, contribute to the electoral loss of the incumbent state government. The forestry practices and the public responses to these that led to the blockade of Giblett constitute the subject of this chapter.

Chinna also examines subsequent effects and actions that have occurred as a result of the Giblett campaign. Based on oral history interviews with protesters and others, Chinna, who was present at the protest, gathers these threads together to memorialize the protest twenty years afterwards. Between Chapters 5 and 6—and between Part 1 "Old-Growth Nature and Culture" and Part 2 "Old-Growth Arts and Activism"—a photographic essay by internationally renowned photographer Juha Tolonen provides illustrations of many of the sites and buildings discussed in *Forest Family* but also offers a rich visual interpretation of old-growth nature and culture in its own right.

Conclusion

As editors, we hope that *Forest Family* will inspire others and provide them with an exemplar for writing similar stories and accounts that integrate family history, natural history of tree species and their forests, environmental history of a biome and bioregion, ecocriticism of writing about the trees and forests, art history of the visual responses to the trees and forests, architectural and heritage history of built structures and conservation history to save and protect the trees and forests. Out of this rich mix of elements, we believe, a multifaceted, transdisciplinary study in the environmental humanities and critical plant studies emerges for a wide range of readers who can find their interests catered for and can broaden them into new and related fields.

Bibliography

Giblett, Rod. 2013. *Black Swan Lake: Life of a Wetland*. Bristol, UK: Intellect Press.
Giblett, Rod. 2006. *Forrestdale: People and Place*. Bassendean, WA: Access Press.
Potapov, Peter, Matthew Hansen, Lars Laestadius, Svetlana Turubanova, Alexey Yaroshenko, Christoph Thies, Wynet Smith, Ilona Zhuravleva, Anna Komarova,

Susan Minnemeyer, and Elena Esipova. "The Last Frontiers of Wilderness: Tracking the Loss of Intact Forest Landscapes from 2000 to 2013." *Science Advances* 3 (1): 1–13. doi: 10.1126/sciadv/1600821.

RBG Kew. 2016. *The State of the World's Plants Report*. Kew: Royal Botanic Gardens, 2016.

Ryan, John. 2012. *Green Sense: The Aesthetics of Plants, Place, and Language*. Oxford: TrueHeart Publishing.

Thoreau, Henry David. 1950. *Walden*. New York: Harper and Brothers.

Winton, Tim. 2017. "The Island Seen and Felt: Some Thoughts about Landscape." Accessed April 30, 2017. https://placesjournal.org/article/the-island-seen-and-felt/.

Winton, Tim. 2015. *Island Home: A Landscape Memoir*. Melbourne: Penguin.

PART 1

Old-Growth Nature and Culture

∵

CHAPTER 2

From Understory to Overstory: Critical Studies of Old-Growth Trees and Forests

John C. Ryan

Ancient forests occupy an especially prominent niche within the human imagining of nature. People from diverse corners of the globe and throughout the ages have imputed various aesthetic, cultural, economic, spiritual, and symbolic meanings to old-growth ecosystems and to the vegetal behemoths contained within them. From the coastal redwoods of Northern California and the kauri of New Zealand's North Island to the monkey puzzle trees of Chile and pedunculate oaks of Białowieża Forest in Poland, such venerable forests are embodiments of time, interlacing human and non-human histories. In their sheer magnitude and obduracy, primordial trees render the past palpable within the immediacy of cross-species encounters within transhuman communities. Indeed, as "living monarchs of the world's forests," old-growth supplies "a tangible link with the past" and "a spiritual ground in the present" (Maser 1994, 205).

Bristlecone pines, the oldest known living trees, germinated during the founding of Troy and have borne witness to an excess of five-thousand winters. What is more, clonal quaking aspen communities range from eighty-thousand to one-million years in age and derive from a single progenitor that emerged during the Upper Paleolithic period, or much earlier (Hall, James, and Baird 2011, 316). Bringing into experiential proximity a living biocultural heritage that vastly exceeds the human cognitive frame, ancient trees confound our comparatively limited temporal and ontological grasp. In these terms, Robert Pogue Harrison (1992) suggests that "it is not only in the modern imagination that forests cast their shadow of primeval antiquity; from the beginning they appeared to our ancestors as archaic, as antecedent to the human world" (1). As *Forest Family* argues in relation to the long-lived eucalypt ecosystems of the Southwest region of Western Australia, the otherness of old-growth has provoked an array of opposite and, sometimes, intensely contradictory responses—from wonderment and a desire to sacralize to horror and an impulse to instrumentalize.

Harrison's assertion regarding the primeval antiquity of forests is detectable in numerous instances from Western culture, history, and literature—one of which is the forest sanctuary of Zeus at Dodona. Neither a passive backdrop to the exploits of divinities nor a rhetorical device in service to myth, the ancient oak grove proves integral to the sky god who dwells in a notably aged specimen and communicates through the rustling of its leaves (Cusack 2011, 44). Composed around 800 BCE, the famous Greek epic the *Odyssey* alludes to the ancient oak grove and its association with the god Jupiter: "Meantime he voyag'd to explore the will / Of Jove, on high Dodona's holy hill" (Homer 1806, 95, lines 363–64). The contemporaneous *Iliad*, similarly, references prophets "who hear, from rustling oaks, thy dark decrees; / And catch the fates, low-whisper'd in the breeze" (Homer 1813, 366, lines 290–91). While the grove of Dodona is sacred and animate in Greek mythology, Dante Alighieri's "gloomy wood" and "savage wildness" are loci of melancholic introspection, as conveyed in the opening of the *Inferno* from the fourteenth century: "In the midway of this our life below, / I found myself within a gloomy wood,—/ No traces left, the path direct to show" (Dante 1833, 3, lines 1–3).

Later, in his encyclopedic *Histoire Naturelle*, the eighteenth-century French naturalist Comte de Buffon (born Georges Louis Leclerc) bifurcated unculti-vated nature (as hideous) from cultivated nature (as beautiful) (Buffon 1797, 338–39). As the essence of uncultivated nature, old treescapes—especially those of the New World—were inexorably ugly, barren, and hostile. In the conclusion to his oeuvre, the naturalist lamented that "the trees deformed, broken, and corrupted, and the seeds which ought to renew and embellish the scene, are choked by surrounding rubbish, and reduced to sterility. Nature, who in other situations, we find adorned with the splendor of youth, has here the appearance of old age and decrepitude" (Buffon 1797, 336). To be certain, these and other historical patterns persist today—notwithstanding permutations—in some of the discourses surrounding primordial trees. While elevated by some as hallowed sanctuaries deserving protection—an ecocentric standpoint elaborated, for instance, in chapter 8 of the current volume—old-growth forests have been typecast by other writers as unproductive "standing" environments and, thus, reduced to capital-driven volumetrics.

From sacred resonance to biospheric integrity, from aesthetic gravitas to pecuniary possibility, contemporary commentators underscore the heteroge-neous—and, at times, irreconcilable—values of old-growth habitats. As dy-namic bicultural spaces with multiple overlapping histories—Indigenous, colonial, post-colonial, contemporary, natural, ecological, and so forth—old forests epitomize the interpenetration of nature and culture. Sociologist Robert Lee (2009) regards old-growth as "perhaps best defined as a symbolic refuge from an increasingly commercialized world" (99). Presenting a counterforce to the logos of economic optimization, primordial forests possess the potential to transform the Western understanding of space and time, therein providing "refuge for the imagination [...] because natural processes are free to function unimpeded by human demands" (Lee 2009, 99). For nature writer Kathleen Dean Moore (2009, 173), furthermore, old-growth facilitates spiritual well-being and engenders a non-appropriative perspective on the natural world as a refuge from the tyranny of mobile phones, vehicular traffic, computer mediation, artificial light, and, even, one's own misguided internal compulsions.

In addition to aesthetic, cultural, and spiritual values, there is the perhaps more obvious and widely-touted ecological and material utility of forests. They supply genetic resources and non-timber products, buffer against floods and erosion, protect wildlife habitat for ecotourism, research, hunting, and other purposes, and sequester carbon with greater efficiency than their younger re-growth counterparts (Beadle, Duff, and Richardson 2009, 339; Wirth, Gleixner, and Heimann 2009, 5). In this sense, ancient forests are narrativized often-times as "tremendous storehouses of biological diversity" and "genetic reservoirs" (Maser 1994, 206). Rather than wilderness cordoned off from human intervention, as Dante and Buffon imply, old-growth can be further conceptualized as "an emergent property of socio-ecological systems [...] created and sustained by the planned and unplanned conjunction of ecological processes and culturally mediated social choices" (Kanowski and Williams 2009, 345). Consequently, there is a gamut of competing descriptors for old-growth, many of which have resounding—and, on occasion, discomfiting—cultural and historical inflections. Such terms vary from ancient, antique, climax, frontier, heritage, indigenous, and intact to late-successional, natural, original, pre-settlement, primary, primeval, pristine, relict, tall open, untouched, and virgin (Wirth et al. 2009, 19).

Contesting Ideas of Old-Growth

The contributors to this volume argue that *old-growth* is a contested concept, jargonized by bureaucrats, enshrined in federal acts, and marginalized, to some extent, by the resources sector. As invoked by forest activists, conservationists, and artists, the term tends to encapsulate the age, size, venerability, rarity, and fragility of climax habitats as well as their metaphysical resonances. Attempting to particularize the idea and decouple it from transcendental excess, ecologists look toward a host of criteria, including structural, successional, and biogeochemical factors (Wirth et al. 2009, 12–18). Structural criteria emphasize the age and size distributions of forests as well as spatial patterns of living and dead trees (12). Succession theory, in contrast, discerns between *autogenic* changes—those caused by plant interactions—and *allogenic* processes of fire, wind, or clear-felling (clear-cutting) but normally excludes the age and size of a stand. The successional approach highlights the processes leading to primordial communities, namely the gradual replacement of species over time (15–16). In comparison to the preceding two frameworks, the biogeochemical model centers on the often-impracticable quantification of phenomenon such as closed nutrient cycles, zero net biomass accretion, and increased understorey vegetation (Wirth et al. 2009, 18).

Notwithstanding these divergent criteria and approaches, most forest ecologists would agree that old-growth generally displays few signs "of past stand-replacing disturbances" and on the whole exhibits a "dynamic equilibrium driven by intrinsic tree population processes" (Wirth, Gleixner, and Heimann 2009, 5). Put differently, ancient forests develop to a great extent without catastrophic episodes of biotic loss and, instead, progress according to the predominantly unimpeded *autopoietic*—or *autogenic*—trajectory of a climax species, such as the karri, marri, and jarrah eucalypts discussed in this volume. Although stable and well-established, old-growth is far from static—as Buffon would have it—but instead reveals a continuum between growth, disturbance, and decay. Primeval forest systems, furthermore, are predicated in particular on nondualistic states of *death-within-life* and *living-while-dying* (see chapters 6 and 7 of this book).

In the 1980s, following the highly visible campaign to block the damming of the Gordon River in western Tasmania and amid the evolution of the environmental movement in response to the expanding resources sector (Hutton and Connors 1999), the idea of *old-growth* began to infiltrate the Australian imagination more fully. The term became associated with mature forests that appear to be independent of modern technological intrusion and that display unique characteristics of antiquity and tranquility (Beadle, Duff, and Richardson 2009, 339). Public debate during these early years of contemporary Australian eco-activism framed old-growth in numerous ways, for instance, as grounded in environmental science and specifying a developmental stage of a forest; as a generalized concept characterizing intact areas dominated by natural processes; and as a social construct privileging the age of a forest and its relative freedom from human interposition (Keenan and Read 2012, 215).

During the last thirty years in particular, climax forests have achieved iconic standing, galvanizing Australian environmental consciousness while, at times, exacerbating community divisions over concerns of economic livelihood, on the one hand, and ecological preservation, on the other (Kanowski and Williams 2009, 341). As discussed in historical studies of sustainability, nonetheless, centuries-old traditions of Western forestry management, for instance, in Germany (Uekötter 2014) and France (Mather, Fairbairn, and Needle 1999), imbricate economic production with ecological stewardship. The preservation versus conservation debate within forestry management, moreover, can be traced back to nineteenth-century American environmentalism (Martinez 2014, 165–70). Bearing this context in mind, I suggest that the National Forest Policy Statement (NFPS) (Commonwealth of Australia 1992) provided for limited protection of the country's primordial treescapes. The policy characterized old-growth as "forest that is ecologically mature and has been subjected to negligible unnatural disturbance such as logging, roading, and clearing" (iii). Old-growth ecosystems, additionally, are those "in which the upper stratum or overstorey is in the late mature to overmature growth phases" (iii). Published five years later, the JANIS Report streamlined the NFPS definition of old-growth as "ecologically mature forest where the effects of disturbances are now negligible," recognized that forests are "not static," and advocated the preservation of a mosaic of age-classes to ensure the continuity of ecological processes with the potential to generate future old-growth (Commonwealth of Australia 1997, 14–15).

The five-yearly *State of the Forests Report* estimates that Australia contains five million hectares (twelve million acres) of recognized old-growth forests with seventy-three percent protected within reserves (Commonwealth of Australia 2013, 6). Only a small percentage of Australian forests, however, has actually been assessed for old-growth characteristics (103). Emphasis within inventories, moreover, is usually placed on tall and well-hydrated forests, such as the karri of Western Australia, mountain ash of Tasmania, shining gum (*E. nitens*) of Victoria, and blue quandong (*Elaeocarpus angustifolius*) of Queensland. In comparison, long-lived semi-arid ecosystems, notably the mallee woodlands and shrubland of southern Australia (Swan and Watharow 2005), including the Great Victoria Desert, have been excluded historically from the discourse of old-growth and, thus, the ambit of conservation.

Lacking cathedralesque eminence and inhabiting the arid interior, out of sight of the metropolitan centers of the country, mallee and related ecological communities tend to fall outside prevailing concepts of ancient forests. The *State of the Forests Report* encodes a certain image of old-growth as consisting of a layered structure with overstorey trees of substantial girth and height, a well-developed composition of understorey plants, dead standing trees, and decaying logs strewn about the forest floor (Commonwealth of Australia 2013, 102). Even so, the descriptor *ecologically mature*—an alternate term for *old-growth* given in the report—denotes forests "displaying a range of structural, function, and compositional attributes associated with the eco-logical processes characteristic of forests in their mature or senescent growth stages" (Commonwealth of Australia 2013, 663). Accordingly, while karri is both ecologically mature and old-growth, mallee and similar arboreal systems would be excluded from the latter category. And, on a related note, about half of Australia's recognized old-growth is located on public land in New South Wales; in Tasmania, forty percent of the forest cover is categorized as primary (Keenan and Read 2012, 217).

In Australia and elsewhere, old-growth is not only a locus of aesthetic ap-preciation, cultural heritage, economic development, and scientific interest but also an incubator of environmental identity, consciousness, and activism (see chapter 8). In the 1990s, the logging of the ancient temperate rainforests of the Pacific Northwest of the United States prompted the formation of the Ancient Forest Movement and the development of non-violent direct-action protest techniques (Satterfield 2002). Many Tasmanian environmental cam-paigns have centered on forest protection. Beginning in 1999, the Styx Valley campaign, for instance, attempted to halt the clear-felling of giant swamp gums (an alternate common name for *E. regnans*) and made use of tree sits, group marches, and guided walks designed to expose the public to the beauty and fragility of old-growth (Tranter 2009, 709). The ongoing efforts to defend Tasmanian forests are dramatized in first-hand narratives such as Anna Krien's *Into the Woods* (2010). In this regard, it essential to contextualize the old-growth forests of Southwest of Western Australia within the colonial legacies of "slow violence" (Nixon 2011) that are active in Western Australian places, including in the old-growth karri and jarrah country delineated in chapter 3.

In the Southwest of Western Australia, the signing of a Regional Forest Agreement (RFA) between Western Australian and Commonwealth govern-ments was met with public outcry (Commonwealth and Western Australian RFA Steering Committee 1998). The RFA was followed by the implementa-tion of a revised policy known as Protecting Our Old-Growth Forests, which in 2001 set aside over one-million hectares of land and committed to reduc-ing old-growth felling significantly (Houghton 2012). The Forest Management Plan of 2004–13 resulted in the formation of new national parks and conserva-tion reserves. The latest version of the plan (2014–23) reiterates the Western Australian government's resolve to sustain ancient forests and retains the old-growth definition set out in the JANIS report (State of Western Australia 2013). Despite its pretenses of terminating old-growth logging in WA, however, the plan has in fact opened the way for the harvesting of six- to seven-thousand hectares of jarrah forest and four- to five-hundred hectares of karri annually (The Wilderness Society 2015). This is due in part to variability in land classifications and other technical loopholes that make possible the clearing of certain primordial treescapes.

To be certain, as the above discussion makes clear, since European settlement of Australia, old-growth has been the object of conflicting interests—from infrastructure development and nature conservation to market trends and site-based tourism. Proponents of economic rationalism and sustainable use often claim that younger regrowth forests are indistinguishable from their ancient analogues (Creutzburg et al. 2017). The liquidation of old-growth for short-term financial gain, however, jeopardizes biospheric function and integrity at local, regional, and international scales. The bleak reality is that forty percent of precolonial tall open-forests (TOFs) has been lost (Dean and Wardell-Johnson 2010). In the Southwest corner of Western Australia, discussed in the next chapter and throughout this volume, only thirty-seven percent of original forests remains (Dean and Wardell-Johnson 2010, 181).

Notwithstanding such environmentally destructive historical patterns, a notably salient function of old-growth in the Anthropocene—the proposed name for the current epoch denoting the extent of human impact on the biosphere—is the mediation of climate change. It is indisputable that primeval forests amass substantial amounts of carbon in both solid and liquid phases (Dean and Wardell-Johnson 2010, 181). Indeed, the greatest contributors to carbon sequestration are large trees. Tasmanian, Victorian, and Southwest Australian forests produce the tallest and bulkiest eucalypts with high capacities to retain carbon from the atmosphere and, as a result, to enhance climate change mitigation (188).

As an example, the stocking of carbon by Victorian mountain ash forests varies from four-hundred tons of carbon per hectare in thirty-year-old stands to one-thousand tons per hectare in two-hundred-and-fifty-year-old stands (Lindenmayer et al. 2015, 60). Old-growth ash forests are known to store more carbon than any other forest ecosystem in the world because of advanced tree ages, wood density, and high amounts of fallen branches and other dead material (Lindenmayer et al. 2015, 71). In comparison to forests in a regenerative stage, old-growth also requires less water and, therefore, competes minimally with other plants, animals, and fungi (Dean and Wardell-Johnson 2010, 189). Old-growth micro-habitats, furthermore, are not present in structurally less complex regrowth forests (189).

Old-Growth as Sacred Topos and Intelligent Being-in-the-World

Writers from diverse traditions and periods have depicted ancient trees as both sacred and hostile, as sources of communion and aversion. An illustration of the former—of sacrosanct communion—is American romantic poet William Cullen Bryant's "A Forest Hymn," first published in 1824. Bryant figures the forest as a consecrated space (Knott 2012, 8) and conveys his esteem for his literary progenitor Wordsworth through a "deistic conception of nature" (Muller 2008, 185–86). For the poet, primordial forests are "God's first temples" and "ancient sanctuaries" where one's ancestors "offered to the Mightiest solemn thanks" (Bryant 1860, 4, 7).

Moreover, the divine is immanent within, not separate to, old-growth nature: "The barky trunks, the ground, / The fresh moist ground, are all instinct with thee" (Bryant 1860, 16). Beyond its visual charisma, attracting the eye to the soaring canopy, the ancient treescape is the multisensorial topos where "that delicate forest flower / With scented breath" (22) draws the eye (and nose) downward toward the understorey and the intricacies of the forest floor. This sensory immanence—founded in the deistic idea of the divine manifesting through the interplay of sensoriality and substance—diverges strikingly from Buffon's revulsion toward, and denigration of, the swampy composition of the earth beneath ancient trees.

From Buffon's standpoint, the ground is "covered with putrid and stagnating water; there miry lands being neither solid nor fluid, are not only impassable, but are entirely useless to the inhabitants of both land and water" (1797, 336–37). In sharp contrast, Bryant alludes to the radical nondualism of primeval ecosystems. The paradox of *death-within-life* is typically enacted not in the sublime heights but in the comparatively accessible lower reaches of the forest: "How on the faltering footsteps of decay / Youth presses—ever gay and beautiful youth / In all its beautiful forms" (24). For the American poet, divine order, which intergrades with ecosystemic order, lends form to human aspiration and being-in-the-world: "In these calm shades, thy milder majesty, / And to the beautiful order of thy works / Learn to conform the order of our lives" (1860, 32). Accompanying Bryant's text, black and white etchings, created by John Augustus Hows and featured in the 1860 edition of the poem, present unpeopled old-growth wilderness and detailed close-ups of forest-dwelling plants.

Alongside the literary, artistic, ecological, and climatic significance of old-growth there is the demonstrated capacity of trees to exchange information among themselves and with other organisms particularly through mycorrhizal networks. In ancient forests, these interspecies communication channels are more deeply engrained because of minimal disruption to underground networks through allogenic factors. Adaptive behavior in trees, such as defense reactions and gene regulation, changes in response to signals received from neighboring trees through mycelial mediation (Gorzelak et al. 2015, 1). Underground "tree talk" facilitated by mycorrhiza—a mutualism between a fungus and plant root—is integral to forest health (1). In his book *The Hidden Life of Trees*, forester Peter Wohlleben (2016, 19) observes that an individual fungus, if unhindered, can occupy several square miles and encompass an entire woodland. In this regard, ecologists have proposed the term *wood-wide web* to refer to the interconnected, symbiotic system formed by trees and mycorrhizal fungi (Simard et al. 1997). Through this playful, yet poignant, neologism, Suzanne Simard and colleagues acknowledge the chthonic network that enormously predates the communication advances of the late-twentieth century and the disembodied modes of human-human interaction made possible through Internet technologies.

Other scientists in the relatively new—and tendentious—field of plant signalling and behavior underscore that communication, learning, adaptive behavior, and self-organization are unequivocal markers of the "mindless mastery" (Trewavas 2002) and non-cerebral form of intelligence peculiar to vegetal life (Trewavas 2016). These contemporary perspectives on the botanical world generally—and old-growth specifically—run counter to Buffon's disconsolate representation of the primordial forest as a "rude mass of coarse herbage [and] trees loaded with parasitical plants, as lichens, algae, and other impure and corrupted fruits" (1797, 336). In all fairness, the French naturalist and his contemporaries could not have been aware that the rudest and most subterranean of the "corrupted fruits" underlies a finely-tuned interdependent structure. Privileging cultivated environments worked over and gentrified by humans, Buffon concluded brazenly and erroneously—in light of evidence emerging today—that "in the uncultivated and desolate regions, there is no road, no communication, and *no vestige of intelligence* [emphasis added]" (1797, 336–37).

Critical Plant Studies and Old-Growth: Some Possibilities

The diverse values and contexts of old-growth presented thus far in this chapter prompt the following questions: what might the consideration of ancient trees and forests contribute to the emergent field of critical plant studies (CPS)? How might critical perspectives broaden our understanding of primordial botanical environments? Indeed, the past few years have marked a upsurge of interest in plants in cultural (Kohn 2013; Ryan 2012; Vieira, Gagliano, and Ryan 2016), literary (Laist 2013), and philosophical (Marder 2013; Stark 2015) areas. These trajectories have been coupled to developments in the science of plant behavior, communication, and intelligence (Gorzelak et al. 2015; Trewavas 2002, 2016; Wohlleben 2016). According to Hannah Stark (2015), critical plant studies challenges "the privileged place of the human in relation to plant life and examines this through a series of lenses: ethical, political, historical, cultural, textual and philosophical" (180).

The interdisciplinary field of critical plant studies highlights the persistence of appropriative attitudes toward the botanical kingdom in which plants tend to be reduced to mere goods—food, oxygen, fiber, medicine, ornament— fulfilling human wants or needs. It goes without saying that all of us require flora for all those purposes. As the contributors to *Forest Family* make clear, humankind is enmeshed in an invariable and inexorable state of symbiotic exchange with the vegetal sphere. The point underscored by critical plant studies, however, is that botanical life is greater than a relatively static material acted upon and (over)worked(over) by mobile creatures. Plants have their own internal modes of being, thinking, and doing that are independent, yet entwined with, human ontology and epistemology (Marder 2013). The critique of liberal humanism at the core of the field calls for a movement away from the metaphorization and aestheticization of plants—as symbols in service to human values or as rarified objects of beauty or sublimity—and toward a plant-inflected reconfiguration of human being-in-the-world (Stark 2015, 185, 194).

In (re)turning attention to botanical being, critical plant studies reverses the trend articulated by scientists, conservationists, and educators as *plant blindness*. Alarmed by the relegation of threatened floristic communities to the outskirts of conservation discourse, biologists have popularized the notion of plant blindness to point out an inclination "among humans to neither notice nor value plants in the environment" (Balding and Williams 2016, 1192). As a propensity to overlook flora, to underestimate its global ecological importance, or to reduce it to appropriable matter, plant blindness could reflect the physiological constraints of the human processing of visual information (Balas and Momsen 2014, 437). Evoking this idea without citing the term, Randy Laist in his introduction to *Plants and Literature* (2013) argues that "it is impossible to overstate the significance of plants to human life, and yet this simple fact is easily overlooked, taken for granted, or, perhaps, actively repressed in the semantic texture of urban, technological consciousness" (10).

Popular discourse tends to construe plants as "a category of things that are alive like we are, but alive in a way that is utterly different, closed off from our capacity for empathy, omnipresent but unknown, seductive but unresponsive" (Laist 2013, 14). While aural, visual, and textual works throughout Western history reveal the symbolic effectiveness of botanical life, especially of flowers, plants are, of course, greater than the sum of their linguistic or symbolic functions. In fact, they possess intrinsic forms of semiosis and meaning-making that are increasingly rendered apparent in laboratory- and field-based studies. To acquire fitness-related benefits, for instance, plants monitor short bursts—or "soliloquies"—of volatile organic compounds (VOCs) released by neighboring species assailed by herbivores or pathogens (Heil and Adame-Álvarez 2010). Botanical volatiles, moreover, have been described as a "language" involving a syntax regulated by "which compounds are produced when, and in what physiological and ecological contexts" (Raguso and Kessler 2017, 28). As these examples indicate, in underestimating vegetal agency, scholars across the sciences and arts have risked limiting the emergence of a particularly vegetal form of knowledge and closing down the potential of critique itself.

Unlike herbs, shrubs, and everyday trees, however, old-growth is comparatively impervious to the processes of backgrounding but, owing to its biomass, more susceptible to instrumentalization. As the different accounts of primordial forests provided in this chapter reveal, old-growth announces itself to the human imagination through an immense spatiotemporal presence. Critical old-growth studies, therefore, would attend to the ontologies of ancient trees and forests in distinction to other botanical environments. Rather than an act of privileging, this would involve acknowledgement of vegetal-human and vegetal-vegetal difference. In view of the charisma of primordial trees, one prospective approach would be a critical aesthetics that "articulates all aspects of aesthetic experience as an interplay between constants in subjectivity (bound up with embodiment), and historical transformation" (Crowther 1993, 20).

The aesthetics of old-growth, thus, would come to reflect embodied experience from a historicized perspective. Aesthetics moreover would embrace a "conservation counter-aesthetics" informed by all five—or more—senses (Giblett 2011, 72). As we know, old-growth engages sight, as one gawks at the overstorey or stands mesmerized by extreme verticality. The cathedral-like effect of ancient forests indeed is redolent of the sublime as the "aesthetically discomfiting" (Giblett 2011, 65). However, as William Cullen Bryant's "A Forest Hymn" (1860) reminds modern readers, ancient ecosystems are also multisensorial theaters where touch, taste, and smell intergrade synergistically with vision and hearing. This sensory combination yields integrative corporeal experience of vegetal nature that protects against forest sublimation. A critical aesthetics of old-growth, additionally, would recognize the grotesquerie and uncanniness of primordial forest systems as functions of ecological adaptation and expressions of climax forest equipoise.

In countering the instrumentalization of old-growth as "standing-reserve" (Heidegger 1977, 17), critical studies would also take into consideration the emplacement of trees at various scales within bioregions. This kind of con-textualization foregrounds the ecological relations of old-growth including its contribution to hydrological cycles. Chapter 3 of *Forest Family* discusses the bioregional milieu of Southwest Australian old-growth. In this regard, botani-cal science demonstrates that the fitness of a plant is "inextricably linked to the specific environment in which it operates" (Trewavas 2016, 542). A critical old-growth studies referent to vegetal percipience would require an understanding of the environment in which the ancient tree has developed; intelligence is connected indissolubly to the place in which it is expressed (Trewavas 2016, 543). More precisely, old-growth vegetal life is enmeshed in two environments, above and below ground, each with certain constraints that require intelligent negotiation, decision-making, and calculated risk-taking. The emplacement of old-growth at multiple levels protects against the impulse to generalize their lives through taxonomic nomenclature, which provides us the names of species but tells us less about the inner (and outer) and the over (and under) worlds of primordial beings. Within the biocultural demarcations of place, a critical studies of old-growth would take form.

Conclusion: Valuing Old-Growth

Across the globe, vestiges of old-growth remain a vital heritage that reminds us of the deep temporality of plants. In an era of escalating pressures on natu-ral populations of trees, a critical approach to old-growth has the potential to catalyze biocultural transformation, thus yielding fresh insights into ancient ecosystems and revealing the extent of our historical and contemporary in-terdependencies with old forests. Encompassing artistic, cultural, ecological, philosophical, and political perspectives, *Forest Family* provokes an array of possible directions for ensuing studies of primordial forest communities in Australia and elsewhere. In particular, the volume underscores the fundamen-tal role of artistic and literary expression to the re-imagining of an environmen-tally-sensible future that foregrounds the diverse capacities (and capabilities) of the plant realm. The contributors to this volume suggest that an essential part of a botanical future is the enduring presence of the world's vegetal giants. Chapter 3 applies these conceptual considerations of old-growth to the trees and forests of the biodiverse Southwest region of Western Australia.

Bibliography

Balas, Benjamin, and Jennifer Momsen. 2014. "Attention 'Blinks' Differently for Plants and Animals." *CBE: Life Sciences Education* 13 (3): 437–43. doi: 10.1187/cbe.14-05-0080.

Balding, Mung, and Kathryn Williams. 2016. "Plant Blindness and the Implications for Plant Conservation." *Conservation Biology* 30 (6): 1192–99. doi: 10.1111/cobi.12738.

Beadle, Chris, Gordon Duff, and Alastair Richardson. 2009. "Old Forests, New Management: The Conservation and Use of Old-Growth Forests in the 21st Century." *Forest Ecology and Management* 258 (4): 339–40. doi: 10.1016/S0378-1127(09)00411-3.

Bryant, William Cullen. 1860. *A Forest Hymn*. New York: W.A. Townsend.

Buffon, Comte de. 1797. *Natural History, Containing a Theory of the Earth, A General History of Man, of the Brute Creation, and of Vegetables, Minerals, etc*. Translated by James Smith Barr. Vol. 10. London: H.D. Symonds.

Commonwealth and Western Australian RFA Steering Committee. 1998. *A Regional Forest Agreement for Western Australia: Comprehensive Regional Assessment*. Bentley, WA: Forests Taskforce Department of the Prime Minister and Cabinet Regional Forest Agreement Steering Committee.

Commonwealth of Australia. 1992. *National Forest Policy Statement: A New Focus for Australia's Forests*. Canberra: Commonwealth of Australia.

Commonwealth of Australia. 1997. *Nationally Agreed Criteria for the Establishment of a Comprehensive, Adequate and Representative Reserve System for Forests in Australia*. Canberra: Commonwealth of Australia.

Commonwealth of Australia. 2013. *Australia's State of the Forests Report*. Canberra: Commonwealth of Australia.

Creutzburg, Megan, Robert Scheller, Melissa Lucash, Stephen LeDuc, and Mark Johnson. 2017. "Forest Management Scenarios in a Changing Climate: Trade-Offs Between Carbon, Timber, and Old Forest." *Ecological Applications* 27 (2): 503–18.

Crowther, Paul. 1993. *Critical Aesthetics and Postmodernism*. Oxford: Clarendon Press.

Cusack, Carole. 2011. *The Sacred Tree: Ancient and Medieval Manifestations*. Newcastle upon Tyne: Cambridge Scholars Publishing.

Dante. 1833. *The Inferno*. Translated by Ichabod Charles Wright. 2 ed. London: Longman, Rees, Orme, Brown, Green and Longman.

Dean, Christopher, and Grant Wardell-Johnson. 2010. "Old-Growth Forests, Carbon and Climate Change: Functions and Management for Tall Open-Forests in Two Hotspots of Temperate Australia." *Plant Biosystems* 144 (1): 180–93. doi: 10.1080/11263500903560751.

Giblett, Rod. 2011. *People and Places of Nature and Culture*. Vol. Intellect Press: Bristol.

Gorzelak, Monika, Amanda Asay, Brian Pickles, and Suzanne Simard. 2015. "Inter-Plant Communication through Mycorrhizal Networks Mediates Complex Adaptive Behaviour in Plant Communities." *AoB Plants* 7 (plv050). doi: 10.1093/aobpla/plv050.

Hall, C. Michael, Michael James, and Tim Baird. 2011. "Forests and Trees as Charismatic Mega-Flora: Implications for Heritage Tourism and Conservation." *Journal of Heritage Tourism* 6 (4): 309–23. doi:10.1080/1743873X.2011.620116.

Harrison, Robert Pogue. 1992. *Forests: The Shadow of Civilization*. Chicago: University of Chicago Press.

Heidegger, Martin. 1977. *The Question Concerning Technology, and Other Essays*. New York: Harper & Row.

Heil, Martin, and Adame-Álvarez. 2010. "Short Signalling Distances Make Plant Communication a Soliloquy." *Biology Letters* 6 (6): 843–45. doi:10.1098/rsbl.2010.0440.

Homer. 1806. *The Odyssey*. Translated by Alexander Pope. Vol. 3. London: J. Johnson et al.

Homer. 1813. *The Iliad*. Translated by Alexander Pope. London: J. Walker et al.

Houghton, D. Stewart. 2012. "Protecting Western Australia's Old-Growth Forests: The Impact of 2001 Policy Changes." *Australian Forestry* 75 (2): 135–42.

Hutton, Drew, and Libby Connors. 1999. *A History of the Australian Environment Movement*. Cambridge, UK: Cambridge University Press.

Kanowski, Peter, and Kathryn Williams. 2009. "The Reality of Imagination: Integrating the Material and Cultural Values of Old Forests." *Forest Ecology and Management* 258 (4): 341–46. doi:10.1016/j.foreco.2009.01.011.

Keenan, Rodney, and Steve Read. 2012. "Assessment and Management of Old-Growth Forests in South Eastern Australia." *Plant Biosystems: An International Journal Dealing with all Aspects of Plant Biology* 146 (1): 214–22. doi:10.1080/11263504.2011.650726.

Knott, John. 2012. *Imagining the Forest: Narratives of Michigan and the Upper Midwest*. Ann Arbor: University of Michigan Press.

Kohn, Eduardo. 2013. *How Forests Think: Toward an Anthropology Beyond the Human*. Berkeley: University of California Press.

Krien, Anna. 2010. *Into the Woods: The Battle for Tasmania's Forests*. Collingwood, Vic: Black Inc.

Laist, Randy. 2013. "Introduction." In *Plants and Literature: Essays in Critical Plant Studies*, edited by Randy Laist, 9–17. Amsterdam: Rodopi.

Lee, Robert. 2009. "Sacred Trees." In *Old Growth in a New World: A Pacific Northwest Icon Reexamined*, edited by Thomas A. Spies and Sally L. Duncan, 95–103. Washington, DC: Island Press.

Lindenmayer, David, David Blair, Lachlan McBurney, and Sam Banks. 2015. *Mountain Ash: Fire, Logging and the Future of Victoria's Giant Forests*. Clayton South, Vic: CSIRO Publishing.

Marder, Michael. 2013. *Plant-Thinking: A Philosophy of Vegetal Life*. New York: Columbia University Press.

Martinez, J. Michael. 2014. *American Environmentalism: Philosophy, History, and Public Policy*. Boca Raton, FL: CRC Press.

Maser, Chris. 1994. *Sustainable Forestry: Philosophy, Science, and Economics*. Vol. St. Lucie Press: Delray Beach, FL.

Mather, A.S., J. Fairbairn, and C.L. Needle. 1999. "The Course and Drivers of the Forest Transition: The Case of France." *Journal of Rural Studies* 15 (1): 65–90.

Moore, Kathleen Dean. 2009. "In the Shadow of the Cedars: Spiritual Values of Old-Growth Forests." In *Old Growth in a New World: A Pacific Northwest Icon Reexamined*, edited by Thomas A. Spies and Sally L. Duncan, 168–75. Washington DC: Island Press.

Muller, Gilbert. 2008. *William Cullen Bryant: Author of America*. Albany: State University of New York Press.

Nixon, Rob. 2011. *Slow Violence and the Environmentalism of the Poor*. Cambridge, MA: Harvard University Press.

Raguso, Robert, and André Kessler. 2017. "Speaking in Chemical Tongues: Decoding the Language of Plant Volatiles." In *The Language of Plants: Science, Philosophy, Literature*, edited by Monica Gagliano, John Ryan and Patricia Vieira, 27–61. Minneapolis, MN: University of Minnesota Press.

Satterfield, Terre. 2002. *Terre Satterfield Anatomy of a Conflict: Identity, Knowledge, and Emotion in Old-Growth Forests*. Vancouver: University of British Columbia Press.

Simard, Suzanne, David Perry, Melanie Jones, David Myrold, Daniel Durall, and Randy Molina. 1997. "Net Transfer of Carbon Between Ectomycorrhizal Tree Species in the Field." *Nature* 388 (6642): 579–82.

Stark, Hannah. 2015. "Deleuze and Critical Plant Studies." In *Deleuze and the Non/Human*, edited by Jon Roffe and Hannah Stark, 180–96. New York: Palgrave Macmillan.

State of Western Australia. 2013. *Forest Management Plan 2014–2023*. Kensington, WA: Conservation Commission of Western Australia.

Swan, Michael, and Simon Watharow. 2005. *Snakes, Lizards and Frogs of the Victorian Mallee*. Collingwood, Vic: CSIRO Publishing.

The Wilderness Society. 2015. "WA's Forests: Time for Another Big Leap Forward!" Accessed April 30, 2017. https://www.wilderness.org.au/articles/wa%E2%80%99s-forests-time-another-big-leap-forward#old-growth-and-hcv.

Tranter, Bruce. 2009. "Leadership and Change in the Tasmanian Environment Movement." *The Leadership Quarterly* 20 (5): 708–23.

Trewavas, Anthony. 2002. "Plant Intelligence: Mindless Mastery." *Nature* 415 (6874): 841. doi: 10.1038/415841a.

Trewavas, Anthony. 2016. "Plant Intelligence: An Overview." *BioScience* 66 (7): 542–51. doi: 10.1093/biosci/biw048.

Uekötter, Frank. 2014. *The Greenest Nation?: A New History of German Environmentalism*. Cambridge, MA: The MIT Press.

Wirth, Christian, Gerd Gleixner, and Martin Heimann. 2009. "Old-Growth Forests: Function, Fate and Value-An Overview." In *Old-Growth Forests: Function, Fate and Value*, edited by Christian Wirth, Gerd Gleixner and Martin Heimann, 3–10. Berlin: Springer.

Wirth, Christian, Christian Messier, Yves Bergeron, Dorothea Frank, and Anja Fankhänel. 2009. "Old-Growth Forest Definitions: A Pragmatic View." In *Old-Growth Forests: Function, Fate and Value*, edited by Christian Wirth, Gerd Gleixner and Martin Heimann, 11–33. Berlin: Springer.

Wohlleben, Peter. 2016. *The Hidden Life of Trees: What They Feel, How They Communicate: Discoveries from a Secret World*. Carlton, VIC: Black Inc.

CHAPTER 3

Forest Giants: Locating Southwest Australian Old-Growth Country

John C. Ryan

The wide-ranging context provided by Chapters 1 and 2 situates *Forest Family: Australian Culture, Art, and Trees*. The old-growth eucalypts of the Southwest of Western Australia are non-human protagonists in the narrative of the Giblett family (Chapters 4 and 5) and the seminal forest protection campaigns of the 1990s (Chapter 8) that swept the region and resulted in models for subsequent environmental activism in Australia. Central to this narrative are karri (*Eucalyptus diversicolor*) and its companion species jarrah (*E. marginata*). Karris cover roughly 200,000 hectares, or 500,000 acres, about one-fifth of which is classified as old-growth. The iconic forests and charismatic trees are limited principally to a high-rainfall coastal strip extending from the towns of Nannup in the north, Augusta in the south-west, and Denmark in the south-east.

The town of Manjimup, where the Gibletts settled, is set within a transitional zone between karri forests to the south and jarrah principally to the north. In comparison, the town of Northcliffe (featured in Chapter 7)—which is fifty-five kilometers, or thirty-four miles, south of Manjimup—is ensconced squarely within *E. diversicolor* territory (see Figure 3.1). The karri belt varies from sixteen to twenty-five kilometers in width but faithfully parallels the Indian Ocean between Albany and Cape Leeuwin. Isolated communities, however, do exist at Mount Many Peaks and the Porongorup Range (Boland et al. 2006, 286). The soils of the old-growth corridor are typically acidic and—to the dismay of colonial-era pastoralists who intuitively correlated large trees to fertile landscapes—deficient in nutrients and trace elements such as zinc, copper, and cobalt (Boland et al. 2006, 286).

Karri usually mixes with marri (*Corymbia calophylla*) and, less frequently, with jarrah, red tingle (*E. jacksonii*), and yellow tingle (*E. guilfoylei*). Common understory species include sheoak (*Allocasuarina decussata*) and karri wattle (*Acacia pentadenia*). What's more, the characteristic growth habit of karri has physiological provenance. The efficient "hydraulic architecture" of cells, transporting water between roots and leaves, enable trees to grow conspicuously tall and straight, unlike the jarrah, marri, tingle, and fellow eucalypt species

© KONINKLIJKE BRILL NV, LEIDEN, 2018 | DOI 10.1163/9789004368651_004

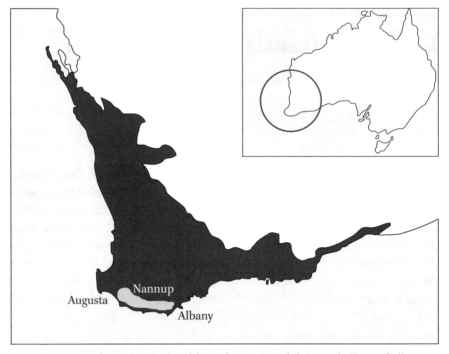

FIGURE 3.1   *Map of Karri Country (grey) located approximately between the Towns of Albany,*
Nannup, and Augusta, Western Australia in the Southwest Botanical Province
(red) (2013).
© MODIFIED VERSION OF ORIGINAL IMAGE FROM PUBLIC DOMAIN VIA
WIKIMEDIA COMMONS.

(McGhee 2014). Alongside the mountain ash (*Eucalyptus regnans*) of Tasmania
and Victoria, the karri ranks as the tallest angiosperm, or flowering plant, in
the world (Griffiths 2001, 17). The largest tree in Australia, in fact, is a mountain ash in Tasmania with a diameter at breast level of more than five meters
and stem volume of three-hundred-and-thirty meters cubed (Beadle, Duff, and
Richardson 2009, 339).

Also known as the Karri Forest Region and the Jarrah-Karri Forest and
Shrublands Ecoregion, the Warren is one of eighty-nine bioregions recognized under the Interim Biogeographic Regionalization for Australia (IBRA).
Demarcated by the Indian Ocean to the south-west and the Jarrah Forest
Bioregion to the north-east, the region has a moderate Mediterranean climate
with the highest rainfalls in Western Australia. The Warren, moreover, is regarded as the most important center of endemism for high-rainfall flora in the
state (Wardell-Johnson and Horwitz 1996).

Southwest Australian Aboriginal Cultures and Old-Growth Forests

The managerialist approach to old-growth codified in policies such as those of the Western Australian government since the 1990s (Chapters 2 and 8) risks eliding the intangible—or less tangible—meanings of primordial forests. Rather than a genetic reservoir, biodiversity storehouse, or object of so-called sustainable yield, old-growth can be regarded as "charismatic megaflora" (Hall, James, and Baird 2011). This idea offers a counterpoint to the economic instrumentalization of old-growth and provides an especially useful optic into the historical (Chapters 4 and 5), artistic and cultural (Chapters 6 and 7), and eco-political (Chapter 8) synthesis of *Forest Family*. In general terms, *charisma* denotes the attractiveness of something and an ability to galvanize a range of relationships, interactions, currents, significations, ideas, and productions.

Geographer Jamie Lorimer (2007) proposes a three-part typology of nonhuman charisma divided along ecological, aesthetic, and corporeal lines. For Lorimer, charisma presents "a bounded relational ontology for considering nonhuman difference" and "draws attention to the importance of affect in understanding environmental ethics" (911). Based on ethnographic research into Southwest forest activism, anthropologists David Trigger and Jane Mulcock (2005) argue that "trees could be said to be *charismatic* in a way that many other plant species are not [emphasis added]" (308). Part of the charisma of old-growth trees, as Trigger and Mulcock suggest, is their ability to mediate spiritual experience and stimulate empathy toward the natural world. Many human visitors to, and dwellers in, karri forests note a sacred immanence that contrasts sharply to the techno-rationalistic and utilitarianist frameworks oftentimes leveraged to the detriment of old-growth well-being (Trigger and Mulcock 2005, 310). While several Anglo-Australian respondents in their study discussed the metaphysical significance of forests in relation to their Christian faith, others intimated some degree of identification with, and interest in, Aboriginal spiritual understandings of old-growth (309).

The Nyoongar are the traditional owners of the landmass extending from the town of Geraldton in the north-west to Esperance in the south-east, and including the Perth metropolitan area (Robertson et al. 2016). Of the fourteen groups comprising the Nyoongar nation, the Pibelmen (or Bibbulmun) are linked traditionally to the high-rainfall old-growth corridor where Manjimup and Northcliffe are located. Drawing from early-twentieth century ethnographer Daisy Bates and contemporary Nyoongar elders, Patricia Crawford and Ian Crawford (2003, 16) apply the term *Murrum* for the Aboriginal people of the Warren Bioregion where Northcliffe is set. Prior to European settlement and, to some extent, during colonial occupation in the nineteenth century, adjacent cultural groups such as the Wardandi, Kaneang, and Minang would have traveled through Pibelmen country to participate in festivals, procuring food and other resources. Following the controversial Regional Forest Agreement (RFA) of the late 1990s (see chapters 2 and 8), Nyoongar people expressed concerns over the limited recognition of the actual extent of Indigenous involvement with old-growth. Stock phrases such as "places of significance" were viewed as obscuring the actual extent of Indigenous relationships to whole ecosystems (McCabe 1998, 5).

For the Nyoongar, and other Australian Aboriginal people, old-growth forests provide material and spiritual sustenance. They are loci of living narratives— of Dreaming stories—in which creation beings underlie the genesis of flora, fauna, fungi, and geological elements. Nyoongar Elder Cliff Humphries, for instance, recalls the story of how the king parrot, or *daarlgayung*, acquired the distinctive red patterns under its wings. In a "terribly tall" karri forest, where animal-beings struggled to ascend the trees but repeatedly fell back to the earth, the king parrot climbed into the canopy and set everything alight with firesticks, or *karl moor* (qtd. in McCabe 1998, 10–11). Old-growth is also associated with the afterlife. If not placed within old trees or stumps, the spirits of the deceased—or *kaarny*—would become *wiriny* or dangerous entities. Inside the tree, however, the placated spirit mediated between worlds, ensuring the success of hunting, harvesting, ceremonial activities, and social dynamics (McCabe 1998, 6). Dennis Eggington summarizes the cultural significance Nyoongar people attribute to ancient forests: "[...] once the old growth is gone and once those special and spiritual places are gone, it is very hard to get the strength from those places and to give the strength back to those places" (qtd. in McCabe 1998, 22).

Old-growth eucalypts figure prominently into the cosmologies, cultural traditions, and artistic expressions of Nyoongar people. The story "The Carers of Everything" dramatizes the actions of Creation Beings who elevate the heavens with the assistance of the branches of tall trees:

> ... they lifted the sky higher and higher. Once it was high enough some of the spirit children turned themselves into coolbardie [magpie], the totem spirit bird for children, and they flew up and pecked the children from the spirit woman's hair and dropped them onto the ground.
>
> NANNUP 2003, 3

The children released from the spirit woman's hair would become the rock outcrops that populate the Southwest landscape. In a different telling of the narrative, a spirit man and woman, who have been appointed the guardians of creation, grow impatient while waiting for *wetj*, the emu. "So they stood up and looked around, and when they stood up they were taller than the giant karri trees and towered over the landscape" (Nannup 2008, 105). The man and woman became arboreal beings who, in their soaring height, like the karris, transcended the physical constraints governing mortal life on earth.

Contemporary poet and playwright Jack Davis (1917–2000) articulated this deep-rooted feeling of Nyoongar people for old-growth trees and forests in his poem "Forest Giant" (2010, 4). Beyond visual appreciation, speculative reverie, and romantic sentimentalization, there is corporeal correspondence between people and trees central to the poem. Davis addresses the old tree directly as a sentient fellow-being: "You have stood there for centuries / arms gaunt reaching for the sky" (2010, 4, lines 1–2). Just as the tree is embodied in the world—its arms reach upward—so too there is the "heart beat of the soil" (line 4) in which the giant grows. Located on an inaccessible slope, and thus sheltered from clear-felling, the tree is a remnant, a survivor who mourns together with the poet: "Now you and I / bleed in sorrow and in silence / for what once had been" (lines 10–12) (see also Chapter 4 for more about Davis' poem).

Echoing the sentiment of Jack Davis, Gaagudju Elder Bill Neidjie is comparably unabashed in his *arborphilia*, or love of trees:

> I love it tree because e love me too.
> E watching me same as you
> tree e working with your body, my body,
> e working with us.
> While you sleep e working.
> Daylight, when you walking around, e work too.
> QTD. IN WINTON 2003, 268

As for Davis, a tree from Neidjie's perspective is a companion who reciprocates emotion, gazes protectively over the world, and works with and beside us. Rather than rendered as other, the forest instead becomes a family member, that is, an intimate relation within a kinship structure based on reciprocal obligations (Clarke 2011).

In contemporary Nyoongar art, old-growth trees also play a vital role in affirming personal identity and recognizing biocultural heritage. As a case in point, Bella Kelly (1915–94) was a Nyoongar artist who was regarded as a formative influence on the Carrolup School, named after the Carrolup Native Settlement near Katanning, Western Australia. Between 1945 and 1951 at Carrolup, Indigenous children of the Stolen Generation produced hundreds of drawings, many of which depicted the local bush. Carrolup, according-ly, became a center of Indigenous modernism after the Second World War (McLean 2016). Spanning five decades but consistently returning to landscape themes, Kelly's artwork often incorporates paddocks and open bush framed by tall jarrahs and karris with the conspicuous Stirling Range looming in the background.

Kelly's paintings are not of an unoccupied and remote wilderness but of *country*—in the Aboriginal sense of the term as ancestral home and nurturer—replete with kangaroos, fences, homesteads, and plentiful traces of past and present human activity. On the ground, broken trees are in vari-ous stages of disrepair and decay. Smaller, understory species such as balga (*Xanthorrhoea preissii*) and tawny patches of soil offer the viewer focal points (Vancouver Arts Centre 2016). In comparison to the grotesque burl that domi-nates the composition of nineteenth-century British botanical artist Marianne North's "Karri Gums near the Warren River, West Australia" (circa. 1880–83) (see Chapter 6), Kelly's work is evocative of the idea of landscape as the inter-mingling of nature and culture—as a place where one lives, works, and recre-ates, as a nourishing terrain of which we are part (Rose 1996) rather than an uncanny scene from which we are alienated.

Colonial-Era Perceptions of Old-Growth

Explorers, cartographers, naturalists, artists, settlers, and, later in the 1800s and early 1900s, venture capitalists were predominantly unaware of the profound cultural significance of ancient forests to Nyoongar people. The newcomers had limited knowledge of the complex Indigenous practices of periodically burning the forest to encourage species richness and modify plant composi-tion. Moist and fertile environments, such as old-growth karri forests, are adapted to low-intensity fires and yield high numbers of herbaceous plants during the period immediately following a burn (Wardell-Johnson et al. 2007). Firing the forest stimulates the grazing of macropods and other animals, which in return enriches hunting. As Bill Gammage observes in *The Biggest Estate on Earth* (2012), landscape burning has served as a form of "planned, precise, fine-grained local caring" (2) for Australian Aboriginal societies for tens of thou-sands of years.

Settler era responses to karri forests, in contrast, mostly foreground the aes-thetic impact and prospective uses of the trees (see also Chapters 4 and 5). In 1831, Captain Thomas Bannister produced one of the earliest written accounts of karri trees. During the exploration of an overland route from Fremantle, near Perth, to King George Sound, near Albany, a distance of approximately four-hundred kilometers (two-hundred-and-fifty miles), Bannister (1980) encountered karris inland from the present-day coastal town of Walpole: "The trees were principally the blue gum; and if others had not seen them, I should be afraid to speak of their magnitude" (105). The explorer invoked an early common designation for the species—*blue gum*—that circulated locally prior to the wider adoption of the Nyoongar name in the late-nineteenth century. Bannister (1980) continues:

> I measured one, it was, breast-high, forty-two feet in circumference; in height, before a branch, 140 or 150 we thought at least, and as straight as the barrel of a gun. (105)

The combination of remarkable stature, striking girth, and the nakedness of the unusually straight and smooth bole prompted Bannister to liken the eucalypt form to a rifle barrel. As with subsequent nineteenth-century chroniclers of the Southwest, many of whom were seeking economic futures centered on the exploitation of natural resources and the conversion of wilderness to settled land, the captain mistakenly deduced that large trees heralded fecund soils: "from the immense growth of these trees, I formed an opinion that the land upon which they grew could not be bad" (Bannister 1980, 105). As Chapter 4 highlights, this presumption—steeped in northern hemispheric consciousness and propelled by colonial boosterism—led to calamitous attempts at agricultural expansion of the region.

The Augusta-area settler John Garrett Bussell was an astute botanical observer, most likely as a result of the length of time he spent working the bush. In May 1830, Bussell took up land near Cape Leeuwin, but, confronted by the labor-intensive task of clearing karri forest there, relocated to a site near the Blackwood River. In his journal of an expedition from the Blackwood to the Vasse River, written sometime in 1832, Bussell (1980) expressed astonishment over the immense size of karris, "that which I have hitherto termed the white gum, (a tree growing to a greater height and bulk than I have yet seen)" (196). While Bannister used the denomination *blue gum*—the modern name referring to species such as *Eucalyptus globulus*, the Tasmanian blue gum—Bussell, in contrast, applied the term *white gum*. This kind of variability between colloquial names for Australian flora would persist until the late 1800s but—in relation to certain trees and wildflowers—continued through the twentieth century and is certainly evident in the present era.

Bussell (1980) metaphorized the unusual process of decortication—the shedding of bark rather than leaves—as the karri "throws off its bark in large flakes, wearing immediately after its change of dress a light buff color" (196). The settler also observed that the karri, for the most part, inhabits "land abounding in springs, having its wood tinged with a light pink" (196). Notwithstanding its aesthetic charisma, the wood proved too dense "for the uses of carpentering, when the eycaliptus [sic] robusta can be obtained" (Bussell 1980, 196). In all likelihood, the ambiguous species mentioned by Bussell is jarrah or Swan River mahogany (*E. marginata*) (Chapter 4 of this volume compares the workability of karri and jarrah timbers). Like Bussell, other writers would highlight the unsuitability of karri for the demands of woodworking. Thomas Laslett (1875), for instance, noted "the defect of star-shake," or timber cracks, and the "peculiar blistery appearance of the annual layers" (197). The forester lamented the incongruence between the imposing stature of karri and the inferiority of its wood as an architectural medium: "It is much to be regretted that a tree so noble in its dimensions should prove so disappointing in its character" (Laslett 1875, 197).

Beginning in the 1870s, the Nyoongar *karri*—less frequently spelled *kari*—became increasingly ubiquitous in the historical record. At this time, the standardization of nomenclature was championed by German-Australian botanist Ferdinand von Mueller. The third volume of his *Fragmenta Phytographiæ Australiæ* (1863, 131–32, published originally in Latin) and, later, the *Report on the Forest Resources of Western Australia* (1879, 6) supply the first known scientific accounts of the eucalypt. Recognizing *E. colossea*—an alternate denomination popular outside of Australia during the nineteenth century—Baron von Mueller propounded the specific *diversicolor*. The term denotes the paleness of the underside of the leaves as well as the color range—from pink to buff—of the wood and bark. Enchanted by the height and girth of the tree as well as the durability of its timber, the botanist aligned karri to the exalted Victorian mountain ash: "This gigantic tree has only one single rival on our island-continent, the Eucalyptus amygdalina (var. regnans) of South-East Australia, the grand features of which it completely repeats. Startling accounts of monster specimen trees are on record, and its maximum height is certainly not overestimated at 400 feet" (von Mueller 1879, 6).

From von Mueller's standpoint, the *monstrous* karri—indeed, an embodiment of grotesquerie—ranked as one of the tallest trees in the world alongside the likes of "Wellingtonia-pine" and "Sequoia-pine" (common appellations for *Sequoiadendron giganteum* of California) and large species of the *Pinus* genus of North America (von Mueller 1879, 6). Hence, within a scientific portrait of the eucalypt and through association with American old-growth, we detect undertones of jingoistic identification mediated by the arboreal world. One of the ways in which the botanist's zeal for superlative Australian trees can be understood is as an expression of pride in the fledgling country and of faith in its progress toward nationhood. Like animals, plants are highly political—and politicized—beings (Vieira, Gagliano, and Ryan 2016, xiv–xvii). Simon Schama (1996), moreover, characterizes the big trees of the United States as "botanical correlates of America's heroic nationalism" (187). Unlike Bussell and Bannister, however, von Mueller predicted that karri would become a significant global commodity. Its wood is "elastic and durable [...] has proved valuable for shafts, spokes, felloes [wheel rims], and rails, and is particularly sought for large planks" (von Mueller 1879, 6).

Following von Mueller's publications, karri semantics underwent a gradual transformation on the whole. As knowledge of the species increased, attitudes toward its economic attributes also changed. In *Western Australia: Its History, Progress, Condition, and Prospects* (1870), William Henry Knight commented that "the Karri (*eucalyptus colossea*) is another wood very similar in many respects to the tuart, and grows to an enormous size" (38). Knight used *colossea*, rather than *diversicolor*, and associated the karri with a fellow hulking—though not necessarily colossal—eucalypt endemic to the Southwest region, the tuart (*E. gomphocephala*). Later in the nineteenth century, the botanist Joseph Maiden in his *Useful Native Plants of Australia* (1889) cited the taxonomic synonym *E. colossea* but noted the diminishing usage of alternate colloquial names for the eucalypt "known as 'Karri', but in its native habitat to a limited extent as 'Blue Gum'" (444). Maiden characterized the wood as light-colored and pliable but straight-grained and tough: "A case is on record of a baulk [beam] of this timber which had been exposed in the wash of the tides at Cape Leeuwin for twenty-six years, continuing sound" (1889, 444).

Considering that Cape Leeuwin—where the Indian and Southern Oceans converge—is infamous for its dramatically turbulent waters, Maiden's anecdote is an especially compelling testimony to the robustness of karri timber. An article in *The Times* from 1896, for example, mentions the export of karri timber to London for paving streets, "as its surface is not easily rendered slippery" (qtd. in Morris 2011, 242). By the publication of the American C.H. Sellers' *Eucalyptus: Its History, Growth, and Utilization* (1910), a perceptible shift in the instrumentalization of the tree occurs. Sellers was a "pro-euc" California forester who played a central role in the popularization of the genus in the Western United States (Farmer 2013, 142). In rather sanguine terms, the forester commented that "the foliage is attractive in appearance [...] The wood is straight-grained and makes good lumber, the tall, straight trunks make good masts" (Sellers 1910, 73).

Contemporary Literary Representations of Southwest Old-Growth

The previous overview of the Nyoongar understandings and colonial-era Anglo-European perceptions—in conjunction with the ecological paradigms and contemporary managerialist views of old-growth trees discussed in Chapter 2—disclose the different values attributed to these charismatic species over time. On the whole, an instrumental perspective constructs forests in terms of possession and appropriation—as *family forests* or, for that matter, as state or federal forests—rather than as a collective of agentic beings, that is, as *forest family*. The simple yet potent reversal of word order presents an ecological heuristic. Such a linguistic inversion hints at what poet John Kinsella (2004) calls "a language of non-contradictory, non-violent but effective protest against the destruction of forests and the environment in general" (161).

The distinction between "family forest" and "forest family" also serves as the central optic of the present volume and supplies a framework for approaching the political, cultural, historical, and artistic representations of—and engagements with—karri environments from the nineteenth century to the present. Not reducible to an object consumed in capitalist networks or an etherealized medium fulfilling the modern yearning for contact with primordial nature, old-growth is home and habitat for humans and non-humans. Rather than limited conceptually to resource or reservoir, karri and other old-growth species (jarrah, marri, tingle) are *oikos*. This term reflects the provenance of the modern concept of *ecology* in the ancient Greek word for household. As this section elaborates in regard to examples from Australian literature of the 1900s and 2000s, old-growth occupies a notable position within the Western literary imagination where, to some extent, it finds release from the utilitarianist ideologies dominating colonialism, as argued in the previous section.

Twentieth-century Australian literary works tend to construe ancient ecosystems in terms of spiritual sustenance and carnal revelation but also exhibit increasing concern over preservation and, later in the period, the limits of human exceptionalism. Katharine Susannah Prichard's *Working Bullocks* (1926) is the seminal novel about karri country and the difficulties faced by timber workers of the early 1900s in the Southwest mill towns surrounding Pemberton (Chapter 4 discusses the novel in further detail). Certain passages of the narrative reveal an arresting sense of empathy for the "long grey logs, corpses of the trees [...] as they lay stretched over the tracks, gliding along to the landing at the mill" (Prichard 1926, 221). Foregrounding the trees as embodied fellow-beings, the narrator confers to karris a conscious presence in the forest that, in some measure, counters—or, at least, underscores—their pervasive commodification and defilement: "The slain giants seemed always on the look out for vengeance—in sullen protest against dismemberment, that tearing of their living flesh by saws in the dark interior of the mill" (Prichard 1926, 221).

A movement away from the sentimentalization and spiritualization of old-growth toward corporeal identification inflecting ecological awareness is nascent in Wolfe Seymour Fairbridge's poem "Karri Forest" (1953, 17–18). Born in Perth in 1918, Fairbridge was a marine biologist and poet whose creative work reflects the combination of a scientific training and a "keen ear for music and rhythm" (Moore 1953, vii). Published more than twenty-five years after Prichard's novel, "Karri Forest" demonstrates a comparable appreciation of the arboreal body in the opening lines:

> Listen!
> Listen!
> Do you hear?
> The whispering columns of the sap ... the ear
> To the great bole; that giant pulse, that heart so near.
> FAIRBRIDGE 1953, 17, LINES 1–5

Rather than anatomizing the karri—linguistically subordinating the tree through tropes, for instance, that would attempt to liken its heart, pulse, and lungs to ours—Fairbridge appeals to the symbiotic affinities between old-growth and humankind: "You hear—/ You hear? / It is your own heart's thunder that you hear" (1953, 17, lines 6–8). The karri forest is "maternal flesh and blood" whose "leaves breathe" (1953, 17–18, lines 16, 32).

Australian poetry of the last twenty-five years displays heightened apprehension over old-growth clear-felling alongside intimations of the idea of forest as family. Originally published in 1991, Dorothy Hewett's "The Valley of the Giants" (2010, 142–43) presents an intriguing take on the familial resonances of karri ecosystems explored in *Forest Family*. The poem opens with a sixty-year-old negative:

> In the burnt-out trunk
> In the karri forest
> myself my little sister
> hand in hand.
> HEWETT 2010, 142, LINES 1–4

Their father looms behind the two children like a "wood demon / deep in shadow / growing out of a tree" (2010, 142, lines 14–16). The three human figures and the karri trunk, hollowed by fire, form a scene "rising up out of the litter / on the forest floor" (2010, 142–43, lines 34–5). For the speaker, the karri forest is the living substrate out of which family history emerges—a perspective indeed shared by Rod Giblett in Chapters 4 and 5 of the present volume. In other words, the tree is consanguineous; its senescence provides a temporal reference point, in this instance, for the progression from life to death, youth to maturity, wide-eyed innocence to hard-edged introspection:

> the giant tree fallen down
> the father dead
> the children grown
> the tragic rotting order overthrown.
>
> 2010, 143, LINES 42–45

As Michael Marder elaborates in *Plant-Thinking* (2013, 103), vegetal life temporalizes human awareness. The poem concretizes Marder's suggestion in the image of the fallen karri inflecting the passage of sixty years since the family posed in the burnt-out trunk. Gaps or ruptures—in this instance, the period between the photograph and the speaker's reflection on the past—are frequently required for the progression of time to register in correspondence to the corporeal transformation of vegetal life.

Karri poetry of the present day reveals an overriding hope for an environmentally-just future that would ensure the well-being of human and non-human lives in the region. Many recent lyrical treatments of karris preserve the feeling of enthrallment over the sheer magnitude of the trees that we find throughout the historical record. In this context, the poet and critic John Kinsella configures the human relationship to old-growth as a pressing matter of linguistics. For Kinsella (2004), "the interaction between the linguistically dominant species (humans) and other species increases (usually to the detriment of the non-human), until the linguistic co-dependent [the human] gains total control, total language" (154). Shifting from old-growth instrumentalism and fetishism to symbiosis and biocentrism entails a reconfiguration of language—an imperative achieved, in part, through poetry and a poetic approach to the world. To be certain, the ancient arboreal life of the Southwest is engrained deeply in Kinsella's writing. He recalls:

> As a twelve-year-old, I went to Pemberton on a school camp. Gloucester Tree lookout was closed; we climbed halfway up Diamond Tree. From our piece of the forest we could see the greater forest spread out. Much less is left now.
>
> KINSELLA 2004, 157

The Gloucester and Diamond Trees are former fire-watching posts that have been converted into popular tourist icons. The former is about thirty-five kilometers (or twenty-two miles) south of Manjimup along the Vasse Highway while the latter is only about ten kilometers (six miles) south of the town along what is known as the South Western Highway. These types of attractions enable direct public access to ancient karris and therein promote the recreational value of old-growth. They, nevertheless, risk opening the way for neocolonialist attitudes toward gargantuan trees as objects of conquest while monumentalizing a few privileged specimens.

Kinsella's "Rapture: Karri Forest at Porongurups" from *The Divine Comedy* (2008) takes place not in the old-growth belt where the iconic two trees are located but in the less-frequented enclave north-east of Albany. The trees are:

> tall enough to prop up the sky,
> hold granite walls, support
> a tunnel of air cleaner
> than logic.
>
> KINSELLA 2008

Although lacking deistic insinuations, the poem intuits William Cullen Bryant's synchronization of internal, societal, and natural orders in ancient forests. Another expression of the physical and metaphysical gravitas of old-growth is Caroline Caddy's "Karri Trees" (2007). Structurally mirroring the appearance of tree limbs and foliage, the poem enacts a reversal of colonial triumphalism and concludes, on an optimistic note, that old-growth will ultimately prevail despite colonial histories of devastation: "the karris whose mighty ancestors / paved the streets of London [...] widen their hold on the sky" (Caddy 2007).

Conclusion: Toward Forest *as* Family

Through literature, visual art, colonial-era accounts, and Indigenous narratives, the central idea developed in respect to the old-growth communities of the Southwest of Western Australia—of a multispecies and dialogical *forest family* in comparison to an anthropomorphized and instrumentalized *family forest*—contributes to post-dualistic theorizations of human histories emerging from complex more-than-human engagements and negotiations (Hamilton and Taylor 2017; Head, Atchison, Phillips, and Buckingham 2016; Kirksey 2014; Kohn 2013; Ryan 2017). Donna Haraway (2008, 16), additionally, theorizes the co-constitutive material-semiotic exchanges between humans and non-humans through her neologisms *naturecultures* and *naturalcultural legacies*. For Haraway and other multispecies theorists, it is impossible for cultural and natural phenomena to exist cordoned off from one other. In particular, history must be understood as the dynamic, ongoing interchange between nature and culture. Nonetheless, Nyoongar and other Indigenous naturecultures disclose that an integrative vision of human-non-human relations on the land—that is, of forest *as* family—is not unique to posthuman or post-colonial critiques of British invasion and Aboriginal dispossession.

In considering these diverse cultural perceptions of old-growth, Chapters 1, 2, and 3 have distinguished in broad terms between, on the one hand, the aesthetic and ethical and, on the other, the instrumental and economic. However, our conceptualization of ecology in *Forest Family*—as *oikos* or household—eschews a binary demarcation between *love for* and *use of* the old-growth of Southwest Australia. From our perspective, dendrophilia *can* exist alongside and within *livelihood*, defined as the sustainable actualization of one's material, economic needs from old-growth ecosystems and the non-human domain more generally. The Indigenous, colonial, and postcolonial encounters with primordial trees presented in this chapter and elsewhere in the book suggest a practice of human-plant inquiry attuned to multispecies communities, nature-culture overlays, and intersubjective being(s)-in-the-world (as outlined in Chapter 2). Furthermore, the overall movement from colonial invasion and deforestation to the post-colonial defense and conservation of old-growth forests signifies the need today to actualize ethical relations to Australian old-growth country. The work of re-visioning, we argue, begins with a conception of history and heritage as inherently multispecies and naturalcultural. To this effect, in the two chapters that follow, Giblett family history supplies the vehicle for narrating the lived, embodied experience of settler-colonial society in the Southwest context.

Bibliography

Bannister, Thomas. 1980. "Report of Captain Bannister's Journey to King George's Sound over Land, Feb. 5th, 1831." In *Journals of Several Expeditions Made in Western Australia, During the Years 1829, 1830, 1831, and 1832*, edited by Joseph Cross, 98–109. Nedlands, WA: University of Western Australia Press.

Beadle, Chris, Gordon Duff, and Alastair Richardson. 2009. "Old Forests, New Management: The Conservation and Use of Old-Growth Forests in the 21st Century." *Forest Ecology and Management* 258 (4): 339–40. doi: 10.1016/S0378-1127(09)00411-3.

Boland, D.H., M.I.H. Brooker, G.M. Chippendale, N. Hall, B.P.M. Hyland, R.D. Johnston, D.A. Kleinig, M.W. McDonald, and J.D. Turner. 2006. *Forest Trees of Australia*. 5 ed. Collingwood, VIC: CSIRO Publishing.

Bussell, John. 1980. "Mr. Bussell's Journal of an Expedition to the River Vasse, from the Blackwood." In *Journals of Several Expeditions Made in Western Australia, During the Years 1829, 1830, 1831, and 1832*, edited by Joseph Cross, 186–203. Nedlands, WA: University of Western Australia Press.

Caddy, Caroline. 2007. *Esperance: New and Selected Poems*. Fremantle, WA: Fremantle Press.

Clarke, Philip. 2011. *Aboriginal People and Their Plants*. Dural Delivery Centre, NSW: Rosenberg.

Crawford, Patricia, and Ian Crawford. 2003. *Contested Country: A History of the Northcliffe Area, Western Australia*. Crawley, WA: University of Western Australia Press.

Davis, Jack. 2010. "Forest Giant." In *Little Book of Trees*, edited by Joanna Karmel, 4. Canberra: National Library of Australia.

Fairbridge, Wolfe. 1953. *Poems*. Sydney: Angus and Robertson.

Farmer, Jared. 2013. *Trees in Paradise: A California History*. New York: W.W. Norton & Company.

Gammage, Bill. 2012. *The Biggest Estate on Earth: How Aborigines Made Australia*. Sydney: Allen and Unwin.

Griffiths, Tom. 2001. *Forests of Ash: An Environmental History*. Cambridge, UK: Cambridge University Press.

Hall, C. Michael, Michael James, and Tim Baird. 2011. "Forests and Trees as Charismatic Mega-Flora: Implications for Heritage Tourism and Conservation." *Journal of Heritage Tourism* 6 (4): 309–23. doi:10.1080/1743873X.2011.620116.

Hamilton, Lindsay, and Nik Taylor. 2017. *Ethnography after Humanism: Power, Politics and Method in Multi-Species Research*. London: Palgrave Macmillan.

Haraway, Donna. 2008. *When Species Meet*. Minneapolis: University of Minnesota Press.

Head, Lesley, Jennifer Atchison, Catherine Phillips, and Kathleen Buckingham, eds. 2016. *Vegetal Politics: Belonging, Practices and Places*. London: Routledge.

Hewett, Dorothy. 2010. *Selected Poems*. Crawley, WA: University of Western Australia Publishing.

Kinsella, John. 2004. "Cross-cut: The Forest." *Antipodes* 18 (2): 153–62.

Kinsella, John. 2008. *Divine Comedy: Journeys Through a Regional Geography: Three New Works*. New York: W.W. Norton.

Kirksey, Eben, ed. 2014. *The Multispecies Salon*. Durham, NC: Duke University Press.

Knight, William Henry. 1870. *Western Australia: Its History, Progress, Condition and Prospects; and Its Advantages as a Field for Emigration*. Perth, WA: J. Mitchell.

Kohn, Eduardo. 2013. *How Forests Think: Toward an Anthropology Beyond the Human*. Berkeley, CA: University of California Press.

Laslett, Thomas. 1875. *Timber and Timber Trees, Native and Foreign*. London: Macmillan and Co.

Lorimer, Jamie. 2007. "Nonhuman Charisma." *Environment and Planning* 25 (5): 911–32. doi: 10.1068/d71j.

Maiden, Joseph. 1889. *The Useful Native Plants of Australia*. London: Trubner and Co.

Marder, Michael. 2013. *Plant-Thinking: A Philosophy of Vegetal Life*. New York: Columbia University Press.

McCabe, Timothy. 1998. *Nyoongar Views on Logging Old Growth Forests*. Perth: Wilderness Society.

McGhee, Karen. 2014. "Searching for Australia's Tallest Trees: Karris." *Australian Geographic*. Accessed April 30, 2017. http://www.australiangeographic.com.au/topics/science-environment/2014/07/australias-tallest-trees-are-karris.

McLean, Ian. 2016. *Rattling Spears: A History of Indigenous Australian Art*. London: Reaktion Books.

Moore, Kathleen Dean. 2009. "In the Shadow of the Cedars: Spiritual Values of Old-Growth Forests." In *Old Growth in a New World: A Pacific Northwest Icon Reexamined*, edited by Thomas A. Spies and Sally L. Duncan, 168–75. Washington DC: Island Press.

Moore, T. Inglis. 1953. "Foreword." In *Poems by Wolfe Fairbridge*, edited by T. Inglis Moore, v–viii. Sydney: Angus and Robertson.

Morris, Edward Ellis. 2011. *Austral English: A Dictionary of Australasian Words, Phrases and Usages*. Cambridge, UK: Cambridge University Press.

Nannup, Noel. 2003. *The Carers of Everything*. Cockburn, WA: Swan Region Strategy for Natural Resource Management.

Nannup, Noel. 2008. "Caring for Everything." In *Heartsick for Country: Stories of Love, Spirit, and Creation*, edited by Sally Morgan, Tjalaminu Mia and Blaze Kwaymullina, 102–14. Fremantle, WA: Fremantle Press.

Prichard, Katharine Susannah. 1926. *Working Bullocks*. London: Jonathan Cape.

Robertson, Francesca, Glen Stasiuk, Noel Nannup, and Stephen Hopper. 2016. "*Ngalak Koora Koora Djinang* (Looking Back Together): A Nyoongar and Scientific Collaborative History of Ancient Nyoongar Boodja." *Australian Aboriginal Studies* 1: 40–54.

Rose, Deborah Bird. 1996. *Nourishing Terrains: Australian Aboriginal Views of Landscape and Wilderness*. Canberra: Australian Heritage Commission.

Ryan, John. 2017. *Plants in Contemporary Poetry: Ecocriticism and the Botanical Imagination*. New York: Routledge.

Schama, Simon. 1996. *Landscape and Memory*. New York: Random House.

Sellers, C.H. 1910. *Eucalyptus: Its History, Growth, and Utilization*. Sacramento: A.J. Johnston.

Trigger, David, and Jane Mulcock. 2005. "Forests as Spiritually Significant Places: Nature, Culture and 'Belonging' in Australia." *The Australian Journal of Anthropology* 16 (3): 306–20.

Vancouver Arts Centre. 2016. "Bella Kelly Retrospective Exhibition." Accessed April 30, 2017. http://www.bellakelly.com.au/introduction.html.

Vieira, Patrícia, Monica Gagliano, and John Ryan. 2016. "Introduction." In *The Green Thread: Dialogues with the Vegetal World*, edited by Patrícia Vieira, Monica Gagliano and John Ryan, ix–xxvii. Lanham, MD: Lexington Books.

Von Mueller, Ferdinand. 1863. *Fragmenta Phytographiæ Australiæ*. Vol. 3. Melbourne: Auctoritate Gubern. Coloniae Victoriæ, Ex Officina Joannis Ferres.

Von Mueller, Ferdinand. 1879. *Report on the Forest Resources of Western Australia*. London: L. Reeve & Co.

Wardell-Johnson, Grant, and Pierre Horwitz. 1996. "Conserving Biodiversity and the Recognition of Heterogeneity in Ancient Landscapes: A Case Study from South-western Australia." *Forest Ecology and Management* 85 (1): 219–38.

Wardell-Johnson, Grant, M.R. Williams, A.E. Mellican, and A. Annells. 2007. "Floristic Patterns and Disturbance History in Karri (*Eucalyptus diversicolor*: Myrtaceae) Forest, South-western Australia: 2. Origin, Growth Form and Fire Response." *Acta Oecologica* 31 (2): 137–50.

Winton, Tim. 2003. "Landing." In *A Place on Earth: An Anthology of Nature Writing From Australia and North America*, edited by Mark Tredinnick, 265–75. Lincoln, NE: University of Nebraska Press.

CHAPTER 4

Family Trees: Jarrah, Karri, and the Gibletts of the Balbarrup-Dingup Area

Rod Giblett

Using pioneering family history from the nineteenth century as a narrative thread woven through the history of, and linking together, the forest and the family, as well as drawing on a range of archival and contemporary sources, such as the cultural and natural history of jarrah and karri trees and forest, Western Australian botanical texts, and management policy for the karri forest, this chapter traces some aspects of the environmental and settler history of the lower Southwest region of Western Australia, centering on the area around the hamlets of Balbarrup and Dingup near the town of Manjimup, and on the Giblett (pronounced with a soft "g" like George) family who were among the first European settlers of this area. Giblett family history is related briefly, and is then placed within the context of the natural and environmental history of the eucalypt jarrah and karri trees and forests of the area. The chapter argues for a dialogics of nature and culture within which the natural environment and non-human context shaped family history and which the family shaped by felling trees, clearing bush, building houses and churches, farming land, and so on. Built and other cultural heritage is discussed in greater detail in the following chapter.

These eucalypt (or gum) tree species and their forests elicited different aesthetic responses and these responses are placed within the context of the history of aesthetics, especially their categories of the sublime, the beautiful, and the picturesque. These trees and forests, especially jarrah, were the environmental basis for a successful timber industry. The psycho-history of the timber industry located in these forests is related by reading ecocritically Katherine Susannah Prichard's classic novel *Working Bullocks* first published in 1926. Prichard insightfully traces the relationships between bush workers (including tree fellers), mill workers, and the Southwestern Australian forest. The Gibletts get a brief mention in this novel (Prichard 1926, 7). As a committed communist, she develops a critique of the ecological politics and political economy of the timber industry. She also touches on the psychological aspects of this resource extraction industry and its investments of desire and capital, yields of pleasure and profit, and relations of work and power. By contrast, no such critique can be found in the tourist propaganda film, *True Australians*, produced in 1948. I go on to critique this film for its jingoistic and triumphalist tale of the Southwest timber industry.

From about 20 years ago, the name "Giblett" came into prominence with the campaign to save the Giblett forest "block" from being logged. This was part of a two-pronged campaign, one a blockade of the forest block and the other a consumer boycott against the timber company Bunnings, now ubiquitous in Australia as the dominant hardware chain store driving small, local hardware stores to the wall by undercutting their prices with sweetheart deals with manufacturers. A gum- or eucalypt-leaf shaped cardboard flyer says that:

> You are called by the Spirit of the Forest to join the Giblett Forest Rescue. Drive to Pemberton. Meet your guide at the Pemberton Tourist Bureau at midday, any day. See and feel the beauty of the forest.

The reverse asks:

> you to help save your old-growth forests and Buy-Pass Bunnings. Sign a pledge not to shop at Bunnings until Bunnings stops clearfelling and woodchipping your native forests.

Nandi Chinna takes up the story of the "Giblett block(ade)" and Southwest forests in Chapter 8.

The Giblett Family

The Gibletts are a pioneering settler family. My great-great grandfather, John Giblett, was born in 1810 and arrived in the Swan River Colony "per [the ship] 'Simon Taylor' in 1842 as an indentured labourer with his wife Ann Wells" who was born in 1808 (Statham 1979, 124). He "was employed first as a labourer in Perth in 1842, then in 1844 as a gardener. By 1846 [he] was working for himself as a tenant farmer" (Statham 1979, 124). He leased Burswood Island in the Swan River for £10 a year to run a dairy. He was "had," or diddled, as this was a swampy marsh, but perhaps the beginning of my association and fascination with wetlands can be traced back to my great-great grandparents. Deacon (1951, 57) describes him as "the tenacious John Giblett" and as "a splendid practical farmer." In 1859 or 1860, the family moved to the Harvey district (Staples

1979, 133). John Giblett built a bridge over Harvey River to reach his farmhouse (Staples 1979, 201).

In 1861 or 1862 they moved to the Manjimup area (Svanberg 1985, 6; Giblett (n.d., 187) says 1861; Deacon (1951, 57) and Berry (1987, 19) say 1862). According to Steward (2008, 3), "the 'Manjimup' property was purchased by John Giblett in 1862 [...] John Giblett could probably be classed as the first permanent resident of Manjimup as he brought a wife and eight children with him to pioneer the land." They were preceded in the shire by three years by the Muirs who were "the first settlers in the Warren District [...] in 1859" (Berry 1987, 16). The Muirs lived further east from the Gibletts around Lake Muir.

Three months, or two, or four years, later the Gibletts moved to Balbarrup to the east of Manjimup. Giblett (n.d., 187) says three months; Steward (2008, 4) says John Giblett "took up additional land at Balbarrup in 1864;" Deacon (1951, 57) says four years; Svanberg does not give a precise time frame, except to say that a house was being built for the family at Balbarrup in the meantime. Giblett's three months or Steward's two years are thus more plausible than Deacon's four years. Ronald Giblett (n.d., 187) relates how "the land was heavily timbered in the early days."

The Gibletts cleared jarrah forest, established farming properties, and built a variety of buildings, some of which still survive today as heritage-listed structures, such as Dingup House and Dingup Church. Many members of the Giblett family are buried or interred in the two Balbarrup cemeteries, including the older pioneer one. In commemoration of their pioneering role in the area, the main street of Manjimup is named Giblett Street after them.

John and Ann Giblett's first son Thomas was born in 1842 and purchased land at Dingup in 1868. He built the first section of Dingup House in 1870 comprising three rooms. This house was gradually expanded into a rambling, L-shape building of seventeen rooms (Steward 2008, 4) (see figure 4.1).

John and Ann Giblett's ninth and second last child was Amos Wells Giblett, my great-grandfather, who was born in 1862 in Manjimup ("Extract from Birth Entry," registration number 6496/62, Western Australia Registrar General). Dingup House still stands to this day and, given the number of bedrooms, functions as a bed and breakfast. The Gibletts were active builders in the local area. Thomas built the nearby stone Dingup Church in 1895, 1896, or 1897 that also still stands to this day (Berry (1987, 31) says 1896). The Centenary of the Church was celebrated in 1995. Thomas also built the "Stone House" at Quanuup in about 1890 (Ipsen 2000, 54). John, Ann, Thomas, and other Gibletts are buried in the dilapidated pioneer Balbarrup Cemetery on Perup Road. As this cemetery is located near Balbarrup Brook, it was thought that the decomposing

FIGURE 4.1 *Juha Tolonen. "Dingup House." 2014. Photograph.*
            IMAGE COURTESY OF THE ARTIST.

corpses were polluting the water so the cemetery was closed. All these instances of surviving Giblett built cultural heritage are discussed in greater detail in chapter 5.

The new Balbarrup Cemetery was opened on the top of a hill closer to Manjimup. My father's ashes are interred there. John Giblett became the inaugural postmaster of the Balbarrup Post Office in 1864 on an annual salary of £5. Amos Giblett ran the post office for some years until he moved to Bridgetown. The Post Office ran for nearly one-hundred years until it was closed in 1964. At the time, it "was noted as being the oldest postal service on location in WA"

(Svanberg 1985, 23). A plaque commemorates the location of the long-lost Post Office on the opposite side of Balbarrup Brook from the pioneer cemetery located at the corner of Balbarrup and Perup Roads. A photograph taken in about 1908 showing Amos Giblett standing outside the Post Office with a group of customers is displayed on the dining room wall in Dingup House. Copies of this photograph are on sale in the Manjimup Tourist Bureau and Visitors' Centre in Giblett Street in Manjimup. I have one hanging on the wall in my study as I write this chapter.

Balbarrup Post Office was located on the side of a hill close to Balbarrup Brook and would have had a pleasant outlook as well as being located close to the Balbarrup homestead and on the intersection of two major roads. Manjimup is where it is because of its suitable geographical location for the construction of the modern transportational technology of the railway. It only became a town with the arrival, and termination, of the railway. Berry (1987, 37) notes that:

> in 1910, more than 50 years after the district was opened up by early set-
> tlers, the first town in the district began to emerge [...] which was to be-
> come the terminus of the railway. Streets in the townsite were named in
> honor of the 'old and prominent settlers.'

These settlers included the Gibletts after whom Giblett Street is named.

My interest in my family history is not prompted merely by the desire to trace my genealogy (which is easily done for four generations), but also by my interest in environmental history and the desire to set my family history within the context of the non-human natural environment which shaped my family history and which the family shaped by felling trees, clearing bush, building houses and churches, farming land, and so on, in the dialogics of nature and culture. The land they cleared and settled was located on the southern edge of the jarrah forest and the northern edge of the karri forest where the two are in transition and intermingle as the predominant species. Some maps show a clear dividing line between jarrah and karri forests with one map at a scale of 1: 50 showing a line through the town Manjimup whereas others, such as one at a finer scale by F.G. Smith of 1972, show the jarrah and karri forests inter-mingling with no hard and fast or clear dividing line between them (see Beard 1990, 60–61).

The two species have some notable differences. Jarrah is easier to cut down and the soil is more fertile for farming, whereas karri is taller and harder to cut down and the soil is poorer for farming. I like to think that my great-great-grandfather knew that karri soils were agriculturally unproductive and did not venture any further south for this reason. He probably knew that karri trees are a lot harder to fell than jarrah being a bigger tree with harder wood. I also like to think he took the wiser and easier option of clearing a farm in jarrah forest than trying to do so in the soils of "the agriculturally unproductive" karri forest (as I will discuss shortly).

I do not regard the natural environment and non-human context of my early Western Australian family history as the mere background or backdrop against which, or simple site in which, human activity took place and to which it was opposed as an antagonist; rather, I regard the natural environment and non-human context as actors and agents in their own right and in and with their own story, which intersects with my family story in a certain time and place. This is the story of people and place, place and people, in their intertwined and living processes. I retell the story of the place in chronological order and the story of the people as a genealogy in reverse chronological order beginning with the Gibletts and going back to, and ending with, Aboriginal people. The story weaves together past, present, and future.

Jarrah

The eucalypt jarrah and karri forests of the botanical and bio-geographical province of Southwestern Australia are unique. The province extends from Shark Bay to Israelite Bay (Christensen 1992, vi) (see also chapter 6 of this volume). The jarrah forest predominates more in the central areas of this province, whereas the karri forests are found solely in the southern area of the botanical province (Christensen 1992, 7). These two species elicited diametrically opposed views of their aesthetic appeal (or lack of it in the case of jarrah) from the first West Australian conservator of forests, John Ednie-Brown, appointed in 1895. In 1899 Ednie-Brown (1899) commented how:

> Taken as a whole, there is nothing particularly picturesque about the appearance of a Jarrah tree or forest of these. Indeed, the general effect of the species, *en masse*, is dull, somber, and uninteresting to the eye. Except in special spots and localities, the trees are rugged and decidedly inclined to be straggling and branchy. (10)

Similarly George Seddon (1997) writing of this region relates how, for the eye trained in the conventions of the European landscape aesthetic:

[...] that balance of foreground, middle ground and background with composed masses to the right and to the left that make up our inherited sense of the picturesque hardly work in banksia woodland or jarrah forest or indeed in much of Australia. (136)

He does not mention karri forest, which may suffer from the same deficiencies on these or other aesthetic criteria. "Our inherited sense of the picturesque" refers to a culturally constructed and historically contingent European landscape aesthetic formalized and promulgated in the eighteenth and nineteenth centuries by landscape painters, landscape architects, and landscape writers.[1]

If banksia woodland and jarrah forest are not picturesque, then what are they? They may just be downright ugly and monstrous. Among Ednie-Brown's litany of evils that he holds against the jarrah tree or forest are that they are "dull, somber, and uninteresting to the eye" and "rugged." If they are uninteresting to the eye, they may be interesting to the other senses, such as touch and smell, but Ednie-Brown does not discuss them. The jarrah tree and forest may be uninteresting to the eye, but jarrah timber is interesting to the eye as it was noted, and still is, for its rich mahogany-like color. Its first common English name was mahogany and it was first marketed in England under the name of "Swan River Mahogany." Jarrah timber was noted for its qualities of durability and workability. It is interesting to the nose and the hand, to the senses of smell and touch, as its sawdust is fragrant and the texture of its sawn or planed timbers is dense and smooth. It is a fine furniture timber (see Wrigley and Fagg 2010, 160) and makes warm and durable furniture, paneling, and staircases, as I can attest from personal experience. The jarrah tree and forest may lack aesthetic value to the sense of sight but jarrah timber has other values for the other senses.

Jarrah trees may be uninteresting to the eye, but they are certainly interesting to the hip pocket. Jarrah forest lacks aesthetic value, but this was more than compensated for by the value of jarrah trees and timber as a commodity. The jarrah tree for Brooker and Kleinig (1996) is "the most widespread of the commercially important timber trees of Western Australia" (214). Jarrah timber has a number of desirable qualities that Anthony Trollope (1873), the English novelist, noted in 1871 including that:

---

1   Such as, for instance, Anthony Trollope on his travels to Australia in 1871. I discuss Trollope's landscape writings on this area and the Australian "bush" in the following chapter, "Built in the Forest." In the present chapter, I discuss his views of jarrah and karri timber.

The wood is very hard, and impervious to the white ants and to water. It is a question whether any wood has come into man's use which is at the same time so durable and so easily worked. It may be that, after all, the hopes of the West Australian Micawbers will be realized in jarrah-wood. (186)

Mr Micawber is a fictional character in Dickens' novel *David Copperfield* whose favorite apothegm and frequent assertion was that "something will turn up." His name, as Trollope demonstrates, became synonymous with anyone who pins their hopes on, or places their faith in, a future material advantage that may or may not be realized. The *Merriam-Webster Dictionary* defines a Micawber as "one who is poor but lives in optimistic expectation of better fortune." The West Australian Micawbers for Trollope are poor, but live in hope that jarrah timber will mean better fortune. Jarrah trees and forests are only considered for their monetary and use-value and not for their ecological and aesthetic value. They are reduced to "wood," the living tree reduced to dead matter to be worked and sold. "Man" is also reduced to the recipient of nature's bounty and/or divine providence of a wood that is both durable (hard, impervious to white ants and water) and workable (into commodities, into timber). In her corporate history of Bunnings, Jenny Mills (1986) comments how "at this time [in the late nineteenth century] in the [Swan River] colony the forests appeared an inexhaustible bounty to be fully exploited" (6).

Later in the same decade in which Trollope was writing, Baron Ferdinand von Mueller, one of the early prominent botanists in Australian colonies, both Western Australia and Victoria, "eulogized the jarrah" (Mills, 1986, 5). For him, "the durability of its timber is unsurpassed by any kind of tree in any portion of the globe" (cited by Mills, 1986, 6). High praise indeed. For these reasons, as Wrigley and Fagg (2010) put it, "jarrah is one of Australia's best-known timber trees" (112). Despite this prominence, Irene Cunningham (1998, 277) considers fifty-seven tree species in her book on native trees with a separate chapter or section devoted to each, but curiously does not devote one to jarrah and only gives this iconic tree species an occasional passing reference (see, for instance, 279).

Karri

Three pages after his disparaging remarks about the jarrah forest cited earlier Ednie-Brown (1899) turns his attention to the karri tree and comments in glowing terms how:

> There is no doubt that this is the finest and most graceful tree in the Australian forests [...] The trees are almost always of straight growth, and tower skywards for great heights without having even the semblance of a branch. So marked are they in these respects that they look like a mass of upright candles. (13)

No doubt, some eastern Australian foresters and tree lovers would take issue with his opening remark. His description of the karri in strongly vertical terms and his simile connotes the sublime gothic cathedral, based, as it is, on forest groves itself. Along similar lines, Charles Lane Poole who succeeded Ednie-Brown as forest conservator was "delighted with the tall, straight-boled karri trees rising like the columns of a cathedral, and occasionally forming themselves into forest aisles" (cited by Cunningham 1992, 278). The forest grove as a cathedral was a cliché of late nineteenth-century and early twentieth-century aesthetics of forests and Ednie-Brown and Lane Poole duly obliged by repeating it. Lane Poole was also an early proponent of "sustained yield" (Mills, 1986, 69). The jarrah for Ednie-Brown is monstrous, whereas the karri for him is monumental. The number of branches of the jarrah certainly contrasts with the lack of branches of the karri. In 1877 Ferdinand von Mueller exclaimed in terms of the sublime how "the Karri [is] one of the grandest trees of the globe and one of the greatest wonders in the whole creation of plants!" (cited by Cunningham 1992, 269). Von Mueller eulogized the jarrah for the durability of its timber and the karri for its sublime aesthetic qualities, two different criteria for assessing trees, one tied up with commodification, the other with aesthetics.

Rather than the aesthetic category and criterion of the picturesque invoked explicitly by both Ednie-Brown and Seddon in relation to the jarrah tree and forest (and found wanting), and the sublime evoked implicitly by Ednie-Brown in relation to the karri (and found therein), Per Christensen in his management plan for the karri forests invokes the aesthetic category of the beautiful without unpacking the criteria by which the karri might be considered beautiful. For him, "the [karri] southern forests of Western Australia are one of the most important and beautiful areas of Australia" for two reasons, one of which is "the occurrence of one of the world's tallest hardwood forests, the karri forest" (Christensen, 1992, 1). More precisely, for Wrigley and Fagg (2010), the karri is "the second tallest hardwood in the world" (21).

In the late eighteenth century both Edmund Burke and Immanuel Kant contrasted the beautiful and the sublime in similar terms. For Burke in his treatise on the sublime and the beautiful, the beautiful dwells on small objects and pleasing, whereas the sublime dwells on great and terrible objects (Burke

1757, 113). Similarly for Kant (1764, 48) in his early *Observations* "the sublime must always be great; the beautiful can also be small." Using these definitions of, and criteria for, the sublime and the beautiful, the karri forest and tree, and not timber, would be sublime, not beautiful. Cunningham (1998) concludes that "sublime as a living tree, karri was far from sublime as timber" (270) as it rotted, and when it was ring-barked it refused to fall thus becoming ghostly trunks, or, as she puts it, "specters that still haunt" (275) and ghostly memorials to tree-destruction.

The conclusion that the karri forest and tree, and not timber, would be sublime, not beautiful is borne out by considering some other aspects of Burke's distinction between the sublime and beautiful. For him, the sublime has, or is associated with, a broken and rugged surface, whereas the beautiful has, or is associated with, a smooth and polished surface (Burke 1757, 114). The surfaces of the karri tree and forest are broken and rugged, though Brooker and Kleinig (1996, 58) describe the bark of the karri tree as "smooth." Whereas the sublime concerns or represents vast objects for Burke (1757, 124), the beautiful represents comparatively small objects. The karri tree is huge, according to Wrigley and Fagg (2010, 21), measuring up to eighty meters (two-hundred-and-sixty-two feet) high. This is debatable as Irene Cunningham (1998) relates how "sections sent to the Paris Exhibition" in 1900 "were certified as coming from a tree 142 meters tall" (269). The karri forest is also huge as it covers 610,000 hectares, though this is minuscule compared to jarrah forest that predominates in about 3.9 million hectares (Lines 1998, 92, 6). Whereas the sublime is dark and gloomy for Burke (1757, 124), the beautiful is not obscure. The karri forest can be quite dark and gloomy as depicted in Henry Prinsep's painting, "Karri Trees, Manjimup c.1910" (see figure 4.2).

Prinsep's painting may invoke or evoke the sublime; it is certainly not picturesque in terms of the European landscape aesthetic, as we saw previously and as Seddon (1997) relates, comprising that "balance of foreground, middle ground and background with composed masses to the right and to the left that make up our inherited sense of the picturesque" (136). A single karri tree in the center foreground of Prinsep's painting dominates it by partially obscuring the middle ground of the buildings and the background of the forest. In the picturesque mode, as Seddon points out, the composed masses to the left and right frame the picture of the landscape to make it *picture*-esque. In Prinsep's painting, by contrast, the composed masses of the other trees in the foreground to the left and right of the central tree obtrude into the center of the pictorial space and do not stand unobtrusively off to either side. Prinsep's painting is anti-picturesque, if not sublime.

FIGURE 4.2   *Henry Prinsep. "Karri Trees, Manjimup c.1910." Oil on canvas. 36 × 23.8 cm.*
IMAGE COURTESY OF NATIONAL GALLERY OF AUSTRALIA, CANBERRA.
REPRODUCED WITH THE PERMISSION OF THE NATIONAL GALLERY OF
AUSTRALIA, CANBERRA.

Besides the gloominess of the forest and vastness of the trees, the strongly vertical lines of the trees and the "portrait" layout of Prinsep's painting places it in the sublime mode. The painting is partly a sublime *portrait* of the karri tree *in situ*, rather than a landscape painting of the surfaces of the land in "landscape" layout. The context for the portrait of the karri tree is the farm indicated by the buildings in the background and the post and rail fence, that marker of private property, painted in strong horizontal lines. These qualities contrast strongly with the vertical lines of the trees. The carving out of the homestead in "the bush" is an indicator of the settler attempted conquest of nature (as discussed in greater detail in chapter 5). In this painting, the two forces and drives contend in the intersection at right angles of the trees, fence, and buildings. Trees and fence are also interwoven as if farm and forest had achieved some sort of rapprochement, but the fence follows a straight line between the trees with some enclosed and others excluded from the property. The painting is an allegory of farm and forest played out in horizontal and vertical lines and shapes crossing each other at right angles and forming a grille of bars in which farm, fence, and forest mutually imprison each other.

Perhaps one aspect of the painting worth noting is that it makes an acerbic comment on the difficulty of trying to establish a farm in the forest. One of the trees in the mid-ground is a ghostly dead gum that may have been ring-barked to kill it prior to possible felling to create a field or paddock. Besides the problematic view of the aesthetic qualities of the karri (or lack thereof in the case of the jarrah), their features also led to problems with reading the signs of the fertility (or lack of it) in the soil in which they grew. Seddon (1997) remarks:

> Another view derived from European experience was that the bigger the trees, the better the soil, which led to heartbreaking attempts at settlement in the marri country in Western Australia, and the Otways and Strzeleckis in Victoria. (76)

Marri is also found in the lower southwest (Beard 1990, 59 and 61; Berry 1987, 7). To Seddon's examples, karri country in Western Australia could be added. Cassi Plate (2005) suggests in similar terms to Seddon that "the common assumption that the bigger the trees, the better the soil, led to heartbreaking attempts at settlement in the karri country of Western Australia" (266). Moreover, the common assumption was also that the more sublime the tree, the greater the signs of God's bounty and providence. Those settlers who thought so on both counts misread the signs for, as Beard (1990) puts it, "the soils [of the karri forest country] are agriculturally unproductive" (60).

Berry (1987) concurs that:

the early explorers and settlers in the colony of Western Australia tended
to take the height of the trees as an indicator of the fertility of the [soil],
but over time this relationship was found to be somewhat spurious. The
early settlers in the Warren District preferred to avoid the karri country,
perhaps because it was [also] hard to clear. (28)

These early settlers included the Gibletts who perhaps avoided karri country
for both reasons.

Felling the Forest to Feed the Monstrous Mill

To reap the bounty of the jarrah and karri forests, the trees had to be felled and
milled into timber. "Man" had to engage in labor in felling and milling trees to
gain this reward and invest capital in building and operating mills. "Man" abstracted the forest into trees, trees into wood and timber, and labor into surplus
value. In terms of a Marxist "land-and-labor" theory of value, "Man" extracted
surplus value from labor and timber, and timber, in turn, from tree and forest.
This entailed different relationships between capitalist and worker, between
worker and worker, and between worker and tree, and forest. It involved exploitation of workers by capitalists, solidarity between workers, and alienation
of workers from their own work and their own bodies as the instruments of
production.

It also involved different relations between workers and tree, and the forest. The solidarity of the pastoral drover and shepherd and the tree-felling
bush worker with each other and, to some extent, *with* the bush, is different
from the solidarity of the industrial farm worker and mill worker (or perhaps
more precisely the mill itself) with each other and *against* the bush. The solidarity of the manual bush worker could express a sense of sacrality towards
the bush and gaining aesthetic and spiritual satisfaction from it, whereas the
machinery of the mill enacts mastery over the bush, alienation from it and
oral sadism against it. In Katherine Susannah Prichard's 1926 novel *Working
Bullocks* set in the southern forests of Western Australia, Deb "wondered how
the men dared put the long sharp-toothed perpendicular saw [...] through the
body of a log" (180). It appalled her when "the saw [...] sets its cruel teeth in the
raw wood" (180). Worse was to follow when "the steam-driven saws ate through
the wood" (180) and the wood let out screams and moans (180, 182, 184). The
result for the bodies of the logs was "dismemberment, that tearing of their living flesh" (177).

The fear of the machines turning on the mill workers gripped them and invaded their fantasies. Deb "did not wonder that Charley Johansen had 'the horrors' when he was drunk [and] fancied the saws were chasing him with their bright shark's teeth." Even stone-cold sober, "Deb had a horror of machines and the way they ate up everything before them [...] like great devouring monsters" (183 and 187). Industrial machines are figured as oral-sadistic monsters. Prichard contrasts "all that inhuman machinery" in "the dark interior of the mill" (177 and 183) (a possible allusion to William Blake's "dark satanic mills") with "the natural gait of work in the bush" and the life of the trees themselves that she personalizes:

> In the forest the fallers, bullock-drivers and men at the bush landings treated great trees respectfully. They watched trees and logs as though they were animated and at any moment might be expected to crush a man out of existence. In the bush, men were reverent of a great tree. They gathered to utter oaths of admiration, standing off to appraise his stature, do him homage, before bringing him down. They celebrated his measurements and magnificence with yarns, legends of great trees, at crib time, smoking and gossiping dreamily. But in the mills there was no time for rites to appease dead trees. (183–184)

There was no time for any sort of rites at all for the mill was subject to the work discipline of industrial capitalism directed by machines and the mechanical measurement of time regulated by the clock.

By uttering oaths of admiration, Prichard's fellers were following in the footsteps of Henry David Thoreau's old Romans, and not in those of his mid-nineteenth century contemporary Concord farmers:

> I would that our farmers when they cut down a forest felt some of that awe which the old Romans did when they came to thin, or let in the light to, a consecrated grove (*lucum conlucare*), that is, would believe that it is sacred to some god. The Roman made an expiatory offering, and prayed, Whatever god or goddess thou art to whom this grove is sacred, be propitious to me, my family, and children, etc.
>
> THOREAU 1854, 223

Who knows what, if anything, the Giblett farmers felt and what oaths they may have uttered or offering they have given when felling the jarrah and karri trees of Balbarrup and Dingup.

Along similar lines to Thoreau, in the 1920s Walter Benjamin (2009) warned that:

> From the earliest customs of nations it seems to come to us as a warning that in accepting what nature so bountifully provides we should eschew the gesture of greed. For there is nothing of our own we are able to give back. It is fitting therefore, that we should show reverence in the taking, by restoring a portion of everything we receive before taking possession of it as our own. Such reverence finds expression in the ancient custom of *libatio* [libation]. (63–64)

Modern industrial capitalism shows no such qualms about embracing the gesture of greed and takes without giving anything back except ruination.[2]

Using modern machinery, the timber industry for William Lines (1998) turns the forest into the wastelands of warfare as the aftermath of what he calls "the global assault on the forests:"

> the timber had been trucked out, and tractors and bulldozers had knocked over and ripped apart every remaining plant, treated the debris as rubbish and dragged and pushed it into heaps, ready for burning. Elsewhere, holocaust fires had left white ash, charred limbs and blackened stumps and had demolished soil architecture. (110–11)

Like trees, we human beings have limbs and trunks. When human limbs are amputated, they leave stumps; when trees are cut down, the remaining part of their trunks are stumps. This common naming of parts of bodies and trees is not just a metaphorical nicety, but is a stark reminder that human beings are in symbiosis with the oxygen-producing plants of the planet earth.

No such views of trees and forest grace the nationalistic and tourist promotional film *True Australians* made in 1948 for the Western Australian government tourist bureau. This twelve-minute newsreel film, part of "The Golden West" series, was probably a trailer made expressly for screening in cinemas before the main feature film. The narrator begins in stentorian tones describing how:

---

2   For some other similarities between Benjamin and Thoreau as what I call "archivists of the everyday" see Giblett (2009, 153–154; 2018, chapter 8). The similarities between them in their approach and contributions to radical green politics, dark ecology, and eco-cultural studies have been minimally explored. For greed in relation to mining, drawing on Benjamin and Trollope, see Giblett (2011, chapter 9).

These are the true Australians, members of an exclusive Australian family. To scientifically-minded people they are *Eucalyptus diversicolour* and *Eucalyptus marginata*. To you and me, they are Karri and Jarrah, Western Australia's principal hardwoods. Fine specimens of clean-limbed karri and the more commercially important jarrah still exist in the few remaining tracts of virgin forest in the heavier rainfall areas of south-western Australia.

Is cutting them down any way to treat "true Australians, members of an exclusive Australian family," presumably referring to their endemicity?

And not just cutting them down but doing so with sadistic delight?:

With rhythmic swings a sharp axe bites into the sturdy bole and jagged teeth rip across its back. A few brief hours and life has ended; a giant is on its way to the forest floor from which it sprang perhaps six or seven hundred years ago. But timber mills are insatiable and their cry is for logs and yet more logs. Now and then a resounding thud tells that another of the mighty have fallen. A centuries' reign has ended …

for "king karri" in the forest "he has graced for centuries." He is "shorn of his grandeur" and taken "through picturesque forest glades" to "the hungry, waiting mill. Timber is scarce and the mills are hungry." In the mills "logs satisfy whining saws" until only "the useless heart remains." The mills are "supplying an essential need." For what, one wonders? The mill is an orally sadistic mechanical monster that greedily consumes the forest. Whereas Prichard uses this trope critically, *True Australians* uses it triumphantly and uncritically.

This view of the felling of a forest giant can be contrasted with the Nyoongar view of Jack Davis in his poem "Forest Giant" that has not been felled (yet):

You have stood there for centuries
arms gaunt reaching for the sky
your roots in cadence
with the heart beat of the soil
High on the hill, you missed
the faller's axe and saw
But they destroyed the others
down the slope
and on the valley floor
Now you and I
bleed in sorrow and silence

for what once had been
while the rapists still
stride across
and desecrate the land.

CITED IN MCCABE 1997–98, 15

Davis has fellow-feeling for the tree for the heart of both bodies of human and tree beat with the same rhythm of the soil and bleed in the same sorrow and silence for the loss of the other trees.

Davis critiques the rape of the land and Prichard critiques industrial capitalism, whereas *True Australians* invokes social Darwinism and quotes Herbert Spencer to construct the forest as the site of struggle for the survival of the fittest. For *True Australians* young trees are "embarking on a grim struggle" in which "only the fittest will survive" to grow into "a stately forest" with "quiet grace and dignity." *True Australians* takes the sustaining trope of Social Darwinism used to naturalize the law of the urban industrial capitalist "jungle" and applies it to the forest. Even though only the fittest survive on this view, they are still condemned to be cut down to feed the greedy mills. For *True Australians* "virgin forest is fast disappearing but a prosperous milling industry can be sustained by careful planning and management." This "magnificent heritage" must be used wisely so that its "gifts will be retained forever for the good of mankind, God's gift to all posterity. The film concludes with the rhyming exhortation to "use wisely then your steward's power/whose life to them is but a passing hour."

Tracing my family history and the environmental history of the area that they settled in has helped me to have a broader and richer appreciation for the Southwestern botanical province in which I live, in which I lived the vast majority of my life (as did my ancestors), and which sustains my life and livelihood today (as it did theirs in the past). In addition, tracing the natural and cultural history of the two iconic tree species of the area in the jarrah and karri has also given me a broader and richer appreciation for these two non-human beings that have lived in the area for thousands of year before my family arrived and still live in it. Furthermore, tracing all four histories has given me a broader and richer appreciation for the bioregional home habitat of the living earth in which I try to live in bio- and psycho-symbiosis.

Finally, tracing these histories and retelling them has not been a merely nostalgic journey or antiquarian inquiry into the past but an appreciation for what we still have in the present. This orientation to the past and present provides what Raymond Williams called resources for a journey of hope into the future (Williams 1985, 243–69; see also Giblett 2009, 138–54). By not only making a connection and becoming attached to local place and its living processes in the present, but also to them in the past and for the future, we can live richer, more sustainable lives in mutuality in time and space in bioregional home-habitats of the living earth.

In this respect, we white fellas would only be learning to do as Aboriginal people did for tens of thousands of years before our arrival in the bioregions of the Swan Coastal Plain and Warren. In the Warren bioregion, as Berry (1987) points out, "the aborigines worked the land on a 'sustainable yield' basis long before the modern man had even coined the term" (9). To them goes the last, and hopefully lasting, word (see also chapters 3 and 7 of this volume for further discussion of the Nyoongar people of Southwest Australia).

Bibliography

Beard, J.S. 1990. *Plant Life of Western Australia*. Kenthurst: Kangaroo Press.

Benjamin, Walter. 1928. *One-way Street and Other Writings*. Translated by J.A. Underwood. London: Penguin, 2009.

Berry, Christopher. 1987. *The History, Landscape and Heritage of the Warren District*. Perth: National Trust of Australia (W.A.).

Brooker, Ian and David Kleinig. 1996. *Eucalyptus: An Illustrated Guide to Identification*. Port Melbourne: Reed.

Burke, Edmund. 1757. *A Philosophical Enquiry into the Origin of our Ideas of the Sublime and Beautiful*, edited by James T. Boulton. London: Routledge & Kegan Paul, 1958.

Christensen, Per. 1992. *The Karri Forest: Its Conservation, Significance and Management*, Perth: Department of Conservation and Land Management.

Cunningham, Irene. 1998. *The Trees That Were Nature's Gift*. Maylands, WA: Irene Cunningham.

Deacon, J.E. 1951. "Pioneering in the South-West: The Story of Manjimup, Part 1." *Journal and Proceedings of the Western Australian Historical Society* 4.3: 54–66.

Ednie-Brown, John. 1899. *The Forests of Western Australia and their Development, with Plan and Illustrations*. Perth: Perth Printing Works.

Giblett, Rod. 2009. *Landscapes of Culture and Nature*. Basingstoke: Palgrave Macmillan.

Giblett, Rod. 2011. *People and Places of Nature and Culture*. Bristol: Intellect Books.

Giblett, Rod. 2018. *Environmental Humanities and Theologies: Ecoculture, Literature and the Bible*. London: Routledge.

Giblett, Ronald A. n.d. *Eastward to the Avon: Family Documents Collected and Researched*. Forrestfield, WA: Ronald Giblett.

Ipsen, Bill. 2000. *Follow That Bell!* Bunbury, WA: Bill Ipsen.

Kant, Immanuel. 1764. *Observations on the Feeling of the Beautiful and Sublime*. Translated by John T. Goldthwait, Berkeley: University of California Press, 1960.

Lines, William. 1998. *A Long Walk in the Australian Bush*. Sydney: University of New South Wales Press.

McCabe, Timothy. 1998. *Nyoongar Views on Logging Old Growth Forests*. Perth: Wilderness Society.

Mills, Jenny. 1986. *The Timber People: A History of Bunnings Limited*. Perth: Bunnings Limited.

Plate, Cassi. 2005. *Restless Spirits: The Life and Times of a Wandering Artist*. Sydney: Picador.

Prichard, Katherine Susannah. 1926. *Working Bullocks*. London: Angus & Robertson, 1980.

Seddon, George. 1997. *Landprints: Reflections on Place and Landscape*, Cambridge: Cambridge University Press.

Staples, A.C. 1979. *They Made Their Destiny: History of Settlement of the Shire of Harvey 1829–1929*, Harvey, WA: Shire of Harvey.

Statham, Pamela. 1979. *Dictionary of Western Australians: Volume 1: Early Settlers 1829–1850*. Nedlands: University of Western Australia Press.

Steward, John. 2008. *Manjimup and the Warren District: Past and Present*. Manjmup, WA: John Steward.

Svanberg, Colin. 1985. "... And a Young Man's Faith in a New Country: The Gibletts of Balbarrup," *Warren-Blackwood Times*, January 9: 6.

Thoreau, Henry David. 1854 and 1849. *Walden and Civil Disobedience*. New York: Random House, 2014.

Trollope, Anthony. 1873. *Australia, Volumes I and II*. Gloucester: Alan Sutton, 1987.

Williams, Raymond. 1985. *Towards 2000*. Harmondsworth: Penguin.

Wrigley, John and Murray Fagg, 2010. *Eucalypts: A Celebration*. Crows Nest, NSW: Allen & Unwin.

CHAPTER 5

Built in the Forest: A Hamlet History of Giblett Cultural Heritage

Rod Giblett

The early Western Australian Gibletts were the first European settlers and builders in the local Manjimup area in the nineteenth century (as we saw in the previous chapter); the Muirs were the first European settlers further east in the larger Warren district. The Gibletts were prolific builders. They built Dingup House and Dingup Church; Balbarrup House and Balbarrup Post Office; flour mills on Wilgarup River and Channybearup Brook; and the "Stone House" at Quanuup near Lake Jasper. The buildings of Balbarrup and Dingup made up the hamlets of these names. They also built structures further afield, such as "One Tree Bridge" over Donnelly River on Graphite Road twenty kilometers west of Manjimup. Dingup House and Church remain standing and functioning to this day and are heritage-listed. Dingup House dates from circa 1870 and now operates as a bed and breakfast. Dingup Church was built in about 1895, is well-maintained, and open to the public. Balbarrup Post Office was opened in 1864 and is commemorated by a cairn with a plaque on site and by a photo of it available from the Visitor Information Centre in Manjimup. A couple of remnant sections of the original "One Tree Bridge" are displayed near the current traffic bridge over Donnelly River and near One Tree Bridge holiday chalets.

The Giblett builders are buried in the old, pioneer Balbarrup cemetery on Perup Road over the other side of Balbarrup Brook from the Post Office. Principal among them was the serial builder, Thomas Giblett, who built Dingup House with his father and Dingup Church, the "Stone House" at Quanuup near Lake Jasper, near the south coast, and "One Tree Bridge" on the Donnelly River. This chapter traces the history of this cultural heritage and places it in the context of the natural heritage of Southwest forests (discussed in greater detail in the previous chapter). It argues that these buildings, their gardens, and the hamlets they made up were a little bit of England plonked down in the forest in order to both carve out a home in it and protect themselves from the unhomeliness of the forest.

Chapters 4 and 5 suggest that the early Giblett pioneers as yeoman farmers were motivated by two contending desires, one to clear a viable farm in the forest on soils that would sustain it and them (as discussed in the previous chapter), and the other to create a pleasing prospect to protect them from the dangers of the forest and for them to enjoy as strangers in a strange land (as discussed in this chapter). Agricultural imperatives and aesthetic preferences go hand in hand in creating landscape out of land. The landscape would be both productive and pleasing—pleasing because it is productive.[1]

John and Anne Giblett arrived in Western Australia in 1842 and eventually settled for keeps in 1864 in the Balbarrup district. The Balbarrup Post Office was initially located in the rear of the Giblett house at Balbarup and then a separate, smaller building was later built located at the junction of Balbarrup and Perup Roads. It was opened in 1864 when a fortnightly mail service started between Bunbury and Balbarrup. John Giblett acted as postmaster for £5 per year. A cairn with a plaque commemorates the site of the Post Office. It is dedicated "in memory of John and Anne Giblett the first settlers in the Balbarrup district who settled on this property in 1861 and to commemorate the 100th year of continuous service of the Balbarrup Post Office which John Giblett opened in 1864." It closed in 1963, ninety-nine years and nine months later. It was the longest continuous serving post office in Western Australia at the time. An archival photograph of the Balbarrup Post Office taken in about 1908 shows Amos Wells Giblett in the background behind his more fancily dressed customers.

The cairn also marks the site of the original settlement of Balbarrup, which must have been more like a hamlet than a village. The *Manjimup Heritage Trail Plan* (Savage, et al. 2014) states that "in 1902 the first town site in the Shire was gazetted at Balgarrup [*sic*] about 5km east of the current Manjimup post office [...] The original town site of Manjimup was laid out in 1910" (12). This *Plan* does not mention the well-established historical facts that members of the Giblett family were pioneering settlers at Balbarrup and builders of Balbarrup hamlet, as well as many of the other structures discussed in the *Plan* (as we will see shortly).

The *Plan* does not cite Chris Berry's report on the history, landscape, and heritage of the Warren District commissioned for the National Trust of Australia in Western Australia and published in 1987 (see Berry, 1987; Savage, et al. 2014, "Bibliography," 66). Berry mentions the Gibletts, as indicated in the previous chapter. The only mention of the name "Giblett" in the *Plan* are in relation to the "Giblett Homestead (fmr)" (Savage, et al. 2014, 91) and Giblett Street, the main street of Manjimup, but no mention is made in the *Plan* of how the homestead or street came to have these names. The *Plan* writes the

Giblett family out of the history and heritage of the Shire of Manjimup. The Gibletts get short shrift from the writers of the *Plan* and can rightly feel aggrieved. Indeed, the present chapter and the previous one aim to set the record straight on that score. The Muirs fair much better in the *Plan* (Savage, et al. 2014, 12, 34, 35, 66). Why, one wonders? The presence of the Gibletts at Balbarrup and Dingup is incontrovertible and provides a narrative thread to draw the story of these places together.

Balbarrup Post Office was also built five kilometers to the north of Dingup Church on Balbarrup Road. Why they were built so far apart remains a mystery. One possible explanation is that the Church was located roughly halfway between Dingup House and Balbarrup House to make the travelling time and distance equal for both arms of the family tree. Why Dingup Church and Balbarrup Cemetery, near the Post Office and on the other side of Balbarrup Brook on Perup Road, were also so far apart is an even greater mystery, especially as church and cemetery are usually placed side by side in order to expedite the passage of the dear departed from this world after their funeral in the church to the other—hopefully better—world via the cemetery.

Thomas Giblett and his wife Maria Agnes Moulton purchased Nelson location 82 in 1868, which they named "Dingup," presumably based on the local Nyoongar name for the area. John and/or Thomas Giblett built Dingup House in about 1870. It was advertised for sale in 2008 in the prestige property section, "Home Hunt: Your Piece of Australia," in *The Weekend Australian Magazine* (2008):

> Dingup House was built by John Giblett using mudbricks fired on the property. What began as a three-room abode for he [sic] and his wife in 1870 became, by decade's end, a 17–room homestead housing a family of nine children. Today the home is the seat of 18a hectares in fertile jarrah/karri country dotted with original wooden outbuildings. It was comprehensively restored two years ago and operates as a B&B. Five of the eight bedrooms have ensuites [...] $2.1 million. (45)

The term "seat" implies that the Gibletts were landed gentry who might also have had a city seat to go with their country seat, but this was not case. In England, John Giblett had been a yeoman, a farmer with a small freehold parcel of land.

Coming to the Southwest of Western Australia enabled John Giblett to become a farmer with a large leasehold at Nelson locations 31 and 32 of 44,000 acres near what is now Manjimup. Perhaps he had pretensions of rising through the class hierarchy and becoming landed gentry. In some respects, he was like the squatters who came to Australia from lowly class origins "back home" to find themselves elevated up the social scale by virtue of acquiring large parcels of Aboriginal land. In other respects, he was not like them as he was granted leasehold whereas squatters occupied Aboriginal land and were granted leasehold or freehold later by virtue of already possessing the land.

John and Ann Giblett had eleven children. My great grandfather, Amos Wells, was the ninth. Family legend says rooms were added to Dingup House to accommodate each new addition to the family. As a result, Dingup House is a rambling, slightly ramshackle, and sprawling homestead in traditional Australian colonial style. The more than half-a-dozen bedrooms make it ideal for a very congenial and spacious Bed and Breakfast (B&B). The present owners—a master-of-many-trades and his wife—restored and renovated it several years ago to make it suitable for a B&B. The previous lessees—a chef and his wife—ran the B&B. The roof is high-pitched, perhaps a vestige of northern climes to allow snow to slide off the roof in winter. The verandahs are low-browed to prevent the hot Australian sun from baking the house. The ceilings inside are high, making the house cooler in summer, but also colder in winter. Dingup House is the surviving ancestral home, peopled by ghosts of Giblett family past. It is largely ignored and unvalued by members of the present-day Giblett family and by the writers of the *Manjimup Heritage Trail Plan* who include it in their "Tourist Asset Inventories" under "Accommodation" and briefly mention it earlier as "a B&B" (Savage, et al. 2014, 46, 73). Hopefully it will be appreciated and valued more by future Giblett family members and visitors to the Manjimup area. A sense of family history hangs heavily over the place, traced in the built heritage of the house that once was the family home and in the garden of exotic plants.

The garden has a wide array of exotic or non-native trees and bushes, some of which seem quite old and possibly original plantings dating from the time of the building of the house. There are no native trees or bushes around the house. The wind rattles like rain through the large, hand-sized leaves of the small-tooth quaking or trembling aspen (*Populus tremuloides*). The name "trembling" comes from the clacking noises the flat stems make when the slightest breath of wind rustles the serrated leaves and knocks the stems together. In French the species is called "*tremble*." The sound of the trees of this species at Dingup House is much louder and harsher than those I heard in Canada where their sound is softer and quieter, as the air is softer and moister. No doubt the much drier climate of Australia makes the stems harder whereas the wetter climate of Canada makes them softer and so the air in Canada is moister than the dry air of Australia.

When I was in Canada a few years ago, I walked in an avenue of aspens in Alaksen National Wildlife Refuge on Westham Island in British Columbia. I felt that the air had a vegetable softness (as Henry David Thoreau would say), the leaves made a quiet, soughing sound. I felt that I was walking into the moist embrace of the trees, rather than being hit by a strong blast of hot, dry air as in Australia in summertime. I also recall that these trees dot the grounds of the International Taoist Tai Chi Society center near Orangeville in Ontario. I especially recall them growing next to the small old barn converted into a Taoist shrine, hearing that trembling sound and seeing the summer light refracted through the leaves as I meditated inside early in the morning on my periodic visits there.

This aspen is a fast-growing deciduous tree native to the cooler areas of North America. It is also the most widely distributed tree of North America growing in all ten of the Canadian provinces—and so not in its three northern territories—and in all but thirteen states of the United States. On May 13, 2014, it became Utah's state tree. This species is related to the European aspen (*Populus tremula*), a species of poplar native to cool temperate regions of Europe and Asia. Like the range of its North American cousin, the European aspen is one of the most widely distributed trees in the world, with a natural range stretching from the Arctic Circle in Scandinavia to North Africa, and from Britain across most of Europe and north Asia to China and Japan. How and when the small tooth North American quaking or trembling aspen came to grow from seed or sapling at Dingup House is a mystery. Given its provenance in North America, it is perhaps unlikely that it is an original planting at Dingup House.

The nineteenth-century poet John Clare called the European aspen "wind enarmoured," primarily meaning "wind-enamoured," given his idiosyncratic spelling, enamoured of the wind, a lover of the wind and loved by the wind, but also wind enarmoured, armoured against the wind. Clare describes how "the brustling noise" (glossed by Merryn and Raymond Williams as "rustling," but also with the connotation of "brushing") "so mimics fast approaching showers" that it fools the shepherd boy into thinking it is raining and so rushing to shelter (Williams 1986, 238 and 247). The aspens brustling at Dingup House certainly sound like rain falling on leaves.

Thoreau was not fooled by the sound of the aspens, but he was mystified by them when, on May 17, 1860, he relates how:

> Standing in the meadow nearly the early aspen at the island, I hear the first fluttering of leaves—a peculiar sound, at first unaccountable to me. The breeze causes the now fully expanded aspen leaves there to rustle

with a pattering sound, striking on one another. It is much a gentle surge
breaking on shore, or the rippling of waves. This is the first softer music
which the wind draws from the forest, the woods generally being comparatively bare and just bursting into leaf. It was delicious to behold that
dark mass and hear that soft rippling sound.

THOREAU 1962, XIII, 299; SEE ALSO HIGGINS, 2017 192

Thoreau experiences in an exemplary fashion the aspen, its sounds and sight,
in an emplaced, embodied, and multi-sensory engagement with the tree and
the forest via hearing, seeing, and tasting (and even with a synaesthetic experience of seeing (or 'beholding') by tasting ('delicious')).

Other trees growing at Dingup House are various species of cypress. This
is very much an English country garden in its style of plantings and species,
though the layout (design implies too much forethought) of the garden seems
fairly random. The large trees are lined up on the perimeter fence of the home
paddock like household guards of a little bit of England plonked down in the
middle of the Australian bush, the jarrah and karri forest, with the paddocks
beyond screened from view by the evergreen trees and in summer by the deciduous trees. The trees in leaf are a screen both to protect those from within
from what is without, and on which to project their phantasies of an English
country garden and their fears about the Australian bush.[2]

At Dingup, one does not look out from the house, verandahs or garden
(as Mr. and Mrs. Andrews do in Gainsborough's famous painting from their
bench) at the pleasing prospect of the rolling countryside that one owns, but
at a screen of English trees and bushes that largely prevents one from looking
at the Australian bush and forest beyond (see figure 5.1).

This screen cocoons one from the darkness and dangers of the Australian
bush and forest, protects one from the harsh sun and wind of hot summers, as
well as permits the warm sun of cold winters to penetrate the leafless deciduous trees in the cooler and wetter months (see figure 5.2).

The garden extends the bounds of the home habitat further than the house
and provides a safe shell in which to live protected from the Australian bush
and an inside surface on which to project phantasies of home (meaning
England). The home-habitat for the Gibletts of Dingup House was not the bioregion of the jarrah and karri forests of the Southwest of Western Australia.

Perhaps for them these forests had no homely associations, no pleasing aesthetic qualities, just as they did not for their contemporaries, such as Anthony

---

2   For discussions of the Australian bush as a screen on which fears and phantasies in film,
    photography and literature have been projected, see Giblett (2009, chapter 7; 2011, chapter 6).

FIGURE 5.1 *Juha Tolonen. "Dingup Prospect." 2014. Photograph.*
IMAGE COURTESY OF THE ARTIST.

FIGURE 5.2 *Juha Tolonen. "Dingup Garden." 2014. Photograph.*
IMAGE COURTESY OF THE ARTIST.

Trollope, the nineteenth-century English novelist who travelled in Australia in 1871. He commented favorably on the good commodifiable qualities of jarrah timber (as we saw in chapter 4). He also commented unfavorably on the poor aesthetic qualities of the jarrah and karri forest, or "the bush" as he called it following Australian convention. He complained of the area between Albany and Perth (further east from "Giblett country") that "the bush in these parts never develops itself into scenery, never for a moment becomes interesting. There are no mountains, no hills that affect the eye, no vistas through the trees tempting the foot to wander" (1873, II: 296). The bush does not necessarily compose itself into the picturesque and, sometimes, as in this case, it never does.

Even when the bush is composed into the picturesque and pleasing prospect of the park-like, it can become monotonous.[3] This is the stock response that Trollope (1873, II) inevitably trots out:

> The fault of all Australian scenery is its monotony. The eye after a while becomes fatigued with a landscape which at first charmed with its park-like aspect. One never gets out of the trees, and then it rarely happens that water lends its aid to improve the view. As a rule it must be acknowledged that a land of forests is not a land of beauty [...] every lover of nature is a lover of trees. But unceasing trees [...] become a bore, and the traveller begins to remember with regret the open charms of some cultivated plain. (22–23)

A land of forests is not a land of beauty because forests are too big to be beautiful and the beautiful too small to be a forest. But what is more interesting and noteworthy here is the fact that the land is supposed to conform to the dictates of the European landscape aesthetic. The land is ascribed agency and given life but this agency and life are supposed to mean that land and water improve the view for the jaded traveller. How anthropocentric, or more precisely andocentric, or just plain Eurocentric, can you get?

Even when the landscape was picturesque and fulfilled aesthetic requirements, and so was not boring to watch, it was boring to travel through and failed to fulfill the transportational requirement of teleology as Trollope (1873, I) related:

> There arose at last a feeling that go where one might through the forest, one was never going anywhere. It was all picturesque,—for there was

---

3   For a discussion of the picturesque and pleasing prospect of the gentleman's park estate and
     park-like Aboriginal country in Australia, see Giblett (2011, chapter 4).

rocky ground here and there and hills in the distance, and the trees were
not too close for the making of pretty vistas through them,—but it was
all the same. One might ride on, to the right or to the left, or might turn
back, and there was ever the same view […] One seems to ride forever
and to come to nothing, and to relinquish at last the very idea of an object. (191–192)

One therefore loses at last the very idea of being a subject until one becomes
lost and abject, and ceases to be one against the many. Abject is Julia Kristeva's
(1982, 1–2) term for the mediating category between subject and object that
makes both possible. The subject-object relationship is evident in nature aesthetics and landscape writing, as Trollope indicates. The thought of relinquishing the object also presents the danger of relinquishing the subject in the face
of "the bush."

Trollope invokes and defines the famous Australian term, "the bush." He ad-
vised that:

readers who desire to understand anything of Australian life should become acquainted with the technical meaning of the word bush. The bush
is the gum-tree forest, with which so great a part of Australia is covered,
that folk who follow a country life are invariably said to live in the bush.
Squatters who look after their own runs always live in the bush, even
though their sheep are pastured on the plains.

Bush not was only synonymous with the country, but was also associated with
the track through and the paddock in the scrub. Bush is pastoralized scrub in
which sheep and cattle graze. The bush is the frontier; scrub is wilderness.[4]

Overarching bush and scrub for Trollope is the omnibus category of "woodland." Trollope (1873, 11) argues later that:

woodland country in Australia,—and it must be remembered that the
lands occupied are mostly woodland,—is either bush or scrub. Woods
which are open, and passable,—passable at any rate for men on horseback,—are called bush. When the undergrowth becomes thick and mat-
ted so as to be impregnable without an axe, it is scrub. (22)

---

4 For discussions of pastoralism, the frontier and wilderness see Giblett (2011, chapters 5
   and 10).

And when it is impassable to men on horseback it is scrub too. "The bush" was an amalgam of different non-urban landscapes of trees and plains, of frontier and forest (but not of settled agriculture, nor of wetlands that were unsettlable until they were drained).

By clearing the bush and felling trees in the forest to make a space for their homestead, to turn space into place, to transform the immensity of space into a place, into *this* place, the pioneer Gibletts were following in the footsteps of their forebears and participating in the wider project of Western civilization. For Harrison (1992) in his Western cultural history of forests, "western civilization literally cleared its space in the midst of forests" (ix). By doing so, "a sylvan fringe of darkness defined the limits of its cultivation, the margin of its cities [farms, and colonial settlements, I would add], the boundaries of its institutional domain." The Gibletts exemplify the institution of these definitions, margins, and boundaries in the jarrah and karri forests, albeit with the addition of a screen of familiar exotic trees that screened (partially blocking, partially mediating) the view of the unfamiliar native trees and their forest beyond.

The Gibletts were also establishing their family domain by doing so. "The governing institutions of the West," including the family, for Harrison (1992), "originally established themselves in opposition to the forest, which in this respect have been, from the beginning, the first and last victims of civic expansion" (ix). Wetlands might also have a claim to being the first and last victims of civic expansion.[5] Wetlands and jarrah and karri forests (and banksia woodlands) in the Southwest of Western Australia (including in and around Perth) were the first and last victims of civic expansion. The Giblett family established themselves in opposition to the forest by felling and screening it, and by extracting a livelihood from the resources of its soils, rather than conserving it and working symbiotically with it and its Aboriginal owners and inhabitants, two discourses of nature probably not available to them at the time.

The agricultural imperatives and aesthetic preferences of the Giblett family were of their time, though their appreciation for the fertility of the forest soils and the ease, or not, of felling jarrah or karri trees, at least made them sensitive to these aspects of the forest (as argued in the previous chapter). They may have feared the native forest, but they also created a farm in it to live from and a garden in it to love. They may have been arboriphobic, and they certainly remade an English landscape in Australia, but they also showed some appreciation for, and sensitivity to, the forest in locating their farms where they did.

---

5   For a discussion of the fraught relationship between cities and wetlands, see Giblett (2016).

Aboriginal owners and inhabitants made their "abode," as Harrison (1992, 200) translates "*oikos*," the root of "ecology," their home/land, *with* the forest, whereas the Gibletts established their abode and built their walled home *against* the forest. Despite this cultural difference, both demonstrate that all human beings, as Harrison (1992; my emphasis) puts it, "dwell not in nature but in *the relation to nature*" (201). Human beings dwell in what Alexander Wilson (1992) calls "the culture of nature." Yet rather than one unitary or homogenous culture of nature (as Wilson implies and discusses), there is a variety of competing historically and geographically contingent cultures of natures (see Giblett 2011 for the cultures of natures).

Whereas Aboriginal owners and inhabitants dwelt in the first culture of nature as symbionts with nature, settlers dwelt in the second culture of nature as parasites on nature. In Harrison's terms, the Gibletts "planted the family tree in one place and dwelled domestically" in a clearing in a forest (Harrison 1992, 198). They practiced the second culture of nature of agriculture, "a means [...] of domesticating the law of vegetative profusion" with introduced species after clearing the vegetative profusion of the native jarrah and karri trees and forests. The Aboriginal owners and inhabitants planted their kinship with many places and species—plant and animal—and dwelled domestically with the forests, bush and wetlands in their first culture of nature (see chapter 3). They practiced sustenance hunting and gathering, a means of domesticating the law of vegetative profusion of native species, whereas the settlers planted their own exotic plant and tree species—as the Gibletts and the Dingup homestead demonstrate.

Oak trees are dotted around the property. Were these and the European trees transplanted as seedlings or saplings, or did they grow from transported seeds? Eucalypts line the avenue entering the property. They are of much more recent vintage. Archival photos show that the garden was much more formally laid out in the past with wide paths and little lawn. The broad swathes of lawn are of much more recent vintage. The grounds have a tennis court. My father remembered staying here as a child with his uncles and aunties during school holidays and playing tennis with his cousins. He grew up in Collie as his father worked there with the Forestry Department. His father and mother met in Manjimup when she ran a boarding house with her sister who became pregnant by the son of a wealthy pastoralist in Victoria and sent off to Western Australia with £10.

John and Ann Giblett's eldest son, Thomas was born in 1843 in Perth and built the "Stone House" at Quanuup in about 1890 and Dingup Church on Balbarrup Road beginning in 1894. The church was also used as a school, which may account for the fireplace, unusual for a church, to try to keep the building warmer in winter. A plaque outside the church relates how in 1894 Thomas commenced building the church with stone hewn with broad axes and cut with saws from a quarry on the Balbarrup property five kilometers to the north. The stone was transported by bullock dray to the church site.

Perhaps the location of the church roughly half way between Dingup House and the Balbarrup house was chosen so that Thomas' children and his parents would have the same distance to travel to church. The building was completed in 1895 and the Anglican Bishop of Perth, C.O.L. Riley, licensed it for use as a church. In 1897 a burning tree fell on Thomas Giblett and killed him. He was 54 years old. In 1897 or '98, the apse was added. In 1903 Bishop Riley dedicated a brass lectern in the church inscribed "in loving memory" of Thomas. In the same year the church was recorded as being used as a school when the Education Department agreed to pay £2 per annum for a teacher. The teacher lived at Dingup House in a servant's room at the back of the house. I stayed in this room in November 2014 on my last visit there. In 1923 the Bishop of Bunbury consecrated the church as "St Thomas' Church." Its centenary was celebrated in 1995. I wrote in the visitors' book on April 20, 2014 (my brother Noel's birthday) that I am a great grandson of Amos Wells Giblett, younger brother of Thomas, and thanked whoever looked after the church for doing so.

The *Manjimup Heritage Trail Plan* (Savage, et al. 2014) says that the school at "Dingup Anglican Church," (as they call it) Balbarrup is:

> a State Heritage Register listed location and has some significance to a broad audience. This site also has a B&B [Dingup House], which could also be incorporated into the heritage experience, and a bigger story could be told here. (46)

Indeed, a much bigger story could be told, and it is being told here in the present chapter and the previous one. It is the story of the Giblett family and their construction of Dingup House and Church. Members of the Giblett family are the main characters in the bigger story of Balbarrup and Dingup. They give the story a narrative thread and human interest that could be told to residents of, and visitors to, the Manjimup areas. Rather than treating these buildings and sites as separate entities (as the *Plan* does), they could be integrated into a guided tour of Giblett cultural heritage and the story of the Giblett family.

Thomas Giblett built the Church and it is, in fact, called "St Thomas'" in commemoration of his role. The *Plan* does not mention the builder of the church and only mentions that the Church is called St Thomas' in one passing exception in parentheses (Savage, et al. 2014, 23). The tragic story of saint Thomas Giblett, the pioneer builder, and his early death would be of interest to a wider audience than just members of the Giblett family. The *Plan* also indicates that Dingup Anglican Church has a number of listings as "Classified by the National Trust, Register of the National Estate, Municipal Inventory, Aboriginal Heritage Sites Register" (Savage, et al. 2014, 99).

Thomas Giblett also built One Tree Bridge over the Donnelly River on Graphite Road, twenty kilometers west of Manjimup. According to the *Manjimup Heritage Trail Plan*:

> Graphite was discovered near the Donnelly River in 1882. One Tree Bridge was built to cart graphite out of the mine, much of which was shipped to New York. The graphite turned out to be poor quality and this market dried up and the venture failed but not before investors were deceived and swindled out of their money.
>
> SAVAGE, ET AL. 2014, 12

The *Plan* again does not acknowledge the role of Thomas Giblett in building the bridge. I don't know if he or any other Gibletts were involved in the mine, or the swindle.

A couple of remnant sections of the old bridge are stacked up not far from the new traffic bridge. This bridge is an exceedingly dangerous bridge with no provision for walkers and riders who have to cross the river here as part of their journey on the Bibbulman and Mundi Bindi trails from Perth to Albany on the south coast. If I had a million dollars, I would get a pedestrian and bike bridge built next to the new traffic bridge where the old one-tree one once was. It should be called "Thomas Giblett Bridge." The trails pass under the new bridge.

There is no interpretation of the remnants of One Tree Bridge. I seem to remember there being some here in the past, including archival photos. This is karri and marri country. The karris and marris stand tall, seemingly immutable and arranged in no avenues, unless a road has been carved through the forest like an aisle in a cathedral, nor are they lined up like sentinels as are the non-native trees at Dingup House. The sight and smell of the karri forest and the river take me back to my childhood and youth of camping in these parts. Glenoran Pool, two-hundred meters away on the Donnelly River, is a still, silent, and smooth body of black water that reflects the karri trees on the other side of the river with almost mirror-like stillness and precision.

The "four aces" are two kilometers west of One Tree Bridge on Graphite Road. A trail leads from the bridge to them running parallel and just out of sight of the road. The four aces are four karri trees lined up in a row and spaced evenly apart, seemingly arrayed for touristic pleasure. At least that is how they are presented next to a picnic place and interpretative sign shelter that describes how karri trees are one of the biggest and oldest living organisms on earth. They are up to two-hundred years old and can grow over ninety meters tall. A further fifteen-minute loop walk goes past the four aces into a "karri glade." Dotted along the karri glade walk are angled, tee-shaped pipe structures made of two-inch diameter metal pipe looking like a toy telescope through which the viewer sees details at the which the angled pipe has been directed, such as a bole on the side of one stately giant growing out of its side like a mole.

Thomas is buried in the Old Balbarrup Cemetery, the original cemetery in the Manjimup Shire, on Perup Road alongside his father, who died in 1882, and his mother, who died in 1897, two months to the day after Thomas died, perhaps of a broken heart. Other deceased Gibletts and other pioneers are buried in this cemetery. Nothing much has changed in the four years since I was last here. The monumental stones endure as the monument workers intended. My father's ashes are interred in the new Balbarrup cemetery on top of a nearby hill on Perup Road. A marble plaque is inscribed:

> In memory of Jim Giblett
> 2/6/1924–23/7/1983
> Son of Jim and Ettie
> [short for Henrietta]
> Husband of Freda
> Father of
> Rodney, Kaye and Noel

My father wanted his ashes to be interred beside his pioneering forebears in the pioneering Balbarrup cemetery, but by that stage it had been closed and his ashes ended up in the new Balbarrup cemetery. Rumor has it that some people were concerned about decomposing human remains leaching from the old cemetery into Balbarrup Brook and polluting it. The new Balbarrup cemetery was consequently located up the hill, well away from the brook. Sandra's and my son Blake was just over a year old when his paternal grandfather died and so has no memories of him. Our daughter Zoe was three years old and has vague memories of him. I was thirty-two-years-old when my father died. He has been gone for thirty-three years, roughly half my life. The first half of my life was hardly much better with his notable absences. In my will I have requested that my ashes be interred beside my father's in order to be closer to him in death than I was in life, though I am closer to him in death than in life now as I wear his signet ring given to him "from the family" in 1943, before he went off to World War II and fought at Tarakan on Borneo Island to which he returned as a missionary after the war and where I was born.

Tracing this family history and the cultural history of the area that they settled in has helped me to have a broader and richer appreciation for the area. Tracing, as I did in chapter 4, the natural and cultural history of the two iconic tree species of the area in the jarrah and karri has also given me a broader and richer appreciation for these two non-human beings that have lived in the area for thousands of year before the Giblett family arrived and still live in it. In addition, tracing in both chapters the intertwined natural and cultural histories and heritage of the area in the forests, and in the buildings and other structures members of the Giblett family built in the past, has given me a broader and richer appreciation for their survival and surviving remnants in the present and promoted hope for their conservation and interpretation in the future.

Bibliography

Berry, Christopher. 1987. *The History, Landscape and Heritage of the Warren District*. Perth: National Trust of Australia (W.A.).

Giblett, Rod. 2009. *Landscapes of Culture and Nature*. Basingstoke: Palgrave Macmillan.

Giblett, Rod. 2011. *People and Places of Nature and Culture*. Bristol: Intellect Books.

Giblett, Rod. 2016. *Cities and Wetlands: The Return of the Repressed in Nature and Culture*. London: Bloomsbury Press.

Harrison, Robert Pogue. 1992. *Forests: The Shadow of Civilization*. Chicago: University of Chicago Press.

Higgins, Richard. 2017. *Thoreau and the Language of Trees*. Oakland: University of California Press.

"Home Hunt: Your Piece of Australia." 2008. *The Weekend Australian Magazine* February 16–17, 45.

Kristeva, Julia. 1982. *Powers of Horror: An Essay on Abjection*. Translated by Leon Roudiez. New York: Columbia University Press.

Savage, Claire, Ryan Zaknich, Ryan Mossny, Rikki Clarke and Ashleigh Brown. 2014. *Manjimup Heritage Trail Plan*. Manjimup: Shire of Manjimup.

Thoreau, Henry David. 1962. *The Journal of Henry D. Thoreau, Volumes I–XIV*, B. Torrey and F. Allen, eds. New York: Dover.

Trollope, Anthony. 1873. *Australia, Volumes I and II*. Gloucester: Alan Sutton, 1987.

Williams, Merryn and Raymond, eds. 1986. *John Clare: Selected Poetry and Prose*. London: Methuen.

Wilson, Alexander. 1992. *The Culture of Nature: North American Landscape From Disney to the Exxon Valdez*. Cambridge: Blackwell.

Photographic Essay: Let No Man Put Asunder

Juha Tolonen

© KONINKLIJKE BRILL NV, LEIDEN, 2018 | DOI 10.1163/9789004368651_007

PART 2

Old-Growth Arts and Activism

∵

CHAPTER 6

From Burls to Blockades: Artistic Interpretations of Karri Trees and Forests

John C. Ryan

Karri trees and forests have captivated photographers, painters, and other visual artists and writers since the colonial beginnings of Western Australia. The natural charisma of *Eucalyptus diversicolor*—its remarkable size, striking verticality, trunk textures, color patterns—continues to inspire and challenge today's artists attempting to devise vocabularies for translating their perceptions of karris to a creative medium. Whereas historical commentators have been inclined to dismiss the aesthetic virtues of jarrahs and marris, with their wild asymmetries and strange exudations,[1] karris have been extolled in more consistent terms for having classically beautiful qualities: smoothness, sleekness, gracefulness, grandeur, sublimity (see also chapters 3 and 4). As one of the tallest eucalypt species in the world, second only to Victoria and Tasmania's mountain ash (*Eucalyptus regnans*) (Boland et al. 2006, 286), the karri tree— as evident in its earliest written and visual representations—feeds a public environmental imagination that longs for solitude, serenity, and a glimpse of the divine in nature. In stark contrast to the appreciation of karris as inspirational tree-beings, however, late nineteenth- and early-twentieth-century photographs of surveying and logging activities convey a much different story. The perspective on karris, instead, reflects ideas of utilitarianism where massive old-growth trees are resources to be exploited or behemoths to be overcome for the sake of settler progress (Crawford and Crawford 2003; Hutton and Connors 1999, 193–4).

Considering these divergent, and, at times, conflicting, attitudes towards karris, this chapter provides an overview of visual art—from historical to contemporary eras—portraying individual trees or more extensive karri forest communities. Whereas some artworks pivot around ideals of untamed wilderness,

---

1 For example, in the late 1800s, John Ednie-Brown, widely regarded as the first expert on Australian timber, commented disparagingly on the aesthetics of jarrah forests: "Taken as a whole, there is nothing particularly picturesque about the appearance of a Jarrah tree or forest of these. Indeed, the general effect of the species, *en masse*, is dull, sombre, and uninteresting to the eye" (10) (see chapters 4 and 5).

fundamentally excluding human figures or anthropogenic impacts, others il-
lustrate—in close detail—the recreational or techno-industrial histories of the
eucalpyt forests. Owing to the considerably extensive body of paintings, photography, prints, and engravings of karris, my discussion will focus exclusively
on *E. diversicolor*, rather than jarrahs (*E. marginata*), marris (*Corymbia calophylla*), and other indigenous eucalypt tree species found throughout karri country.
In order to narrow the material further, I center for the most part on the karri
forests between Manjimup and Northcliffe, excluding the transitional ecological zone between karri and tingle communities (*E. jacksonii*) as well as the trees
of Walpole, Denmark, Mount Barker, and Albany farther to the south-east. The
geographical area delimited comprises Pemberton, Windy Harbour, Quinninup,
Crowea, Karridale, and other small localities roughly congruent with the historical activities of the Giblett family of Manjimup (see chapters 4 and 5). Within
this selected area, I highlight visual artworks either for their appeal, originality,
skill, or technique, or because the creator is a well-known historical figure—
or an important living persona—in Australian art, literature, conservation, or
ecology. The survey of karri art will be organized according to the following historical divisions: Colonial Era (1829–1901), Early Twentieth Century (1901–45),
Late Twentieth Century (1945–88), and Contemporary Period (1988–present).
Although some of the broader art-historical trends, related to Australian and
world art, will be addressed, the chapter as a whole focuses on reading the artworks, contextualizing the artists, and attempting to elicit the particular perceptions of karris evident in the representations.

   A region of unusually high rainfall located in the extreme south-western
corner of Western Australia, *karri country* roughly occupies a long, narrow
belt between Nannup, Manjimup, Denmark, and Albany consisting of about
300,000 hectares typically 16 to 25 kilometres in width and running parallel
to the Indian Ocean coastline (Boland et al. 2006, 286; and chapter 3 of this
volume). The area encompassing Giblett country is characterized by rolling
terrain and acidic, low-nutrient soils. With an annual median rainfall of 800
millimeters (31 inches), the big gum landscape corresponds approximately
with the so-called High Rainfall Zone—one of three biogeographical zones
within the Southwest Botanical Province. As chapter 3 mentioned, the greater region is an international biodiversity hotspot extending from Shark Bay
in the north-west to Israelite Bay east of Esperance in the south-east, including
the Perth metropolitan area (Hopper 2004). The most recent version of the
Interim Biogeographic Regionalization for Australia (IBRA), first proposed
in the 1990s based on prior models by the botanists John Beard and Stephen
Hopper, divides the High Rainfall Zone into three subregions: Swan Coastal
Plain, Jarrah Forest, and Warren (Commonwealth of Australia). Karri country exists chiefly in the Warren, with the exception of outlier communities at

Many Peaks (east of Albany), Mount Barker, and the Porongorup Range (north and north-east of Albany, respectively).

Colonial Era (1829–1901): Swan River Colony to Australian Federation

The colonial era marked the beginning of dramatic environmental change in the south-west corner of the fledgling Swan River Colony, founded by James Stirling in 1829. Whereas the indigenous Nyoongar people had judiciously modified the environment, especially with their strategic use of burning to cultivate plant and animal foods over their more than fifty-thousand-year history (Hallam 1975; and chapter 3 of this volume), Anglo-European colonists set in motion sweeping impacts that would fundamentally alter the very composition of the natural world they encountered. By the 1850s and '60s, colonists began to look away from the more established settlements at Fremantle and the Swan River and toward heavily forested country where they anticipated new lives and untold fortunes in exchange for backbreaking labor. In 1856, Thomas Muir settled in the present-day Manjimup area to cut timber. In the 1860s, Edward Reveley Brockman and Pemberton Walcott established homesteads in the vicinity of what would later be known as Pemberton, commencing small-scale farming operations that depended on clearing karri country. Despite the determined wholeheartedness of pastoralists, however, the landscape was not so easily interpreted. As the ecologist George Seddon (1997) comments, a groundless principle guiding early activities was "the bigger the trees, the better the soil, which led to heartbreaking attempts at settlement in the karri country" (76). Within this historical context, two notable artists—a prominent botanical painter, Marianne North (1830–1890), and a natural history photographer, Archibald James Campbell (1853–1929)—travelled to the forested region and produced significant representations of karris. In particular, Campbell's work coincided with the advent of photography in Australia and its functional use by early naturalists, such as the South Australian Samuel Albert White (1870–1954) (Jones 2011, 31), to document field observations as part of research expeditions.

The earliest extant painting of *E. diversicolor* is Marianne North's "Karri Gums near the Warren River, West Australia" (circa. 1880–83, oil on board), featuring the vertically-pronounced, variegated boles of karris towering over a cluster of understory trees and a family of emus.[2] Born in England in 1830, the

---

2　Royal Botanic Gardens, Kew. "Painting 782, Karri Gums near the Warren River, West Australia." *Marianne North Online Gallery*, n.d., accessed December 19, 2017, http://www.kew.org/mng/gallery/782.html.

nomadic landscape painter, botanical artist, and plant discoverer arrived in 1880 at King George Sound in Albany, WA, by steamer from Melbourne. During her stay in Albany, she met the renowned Western Australian plant illustrator Ellis Rowan, "the flower-painter I had heard so much of" (North 1894, 148). By horse-drawn carriage, North then continued overland to karri country—an arduous, slow, and undoubtedly jarring journey during her time—arriving at "The Warren House," the farm of Edward Revely Brockman near Pemberton (Lane 2015, 210). North's recollections of the farm testify to the process of clearing karri country that had already begun well before her arrival. She describes the homestead as "a rambling untidy house with farm-buildings [...] near a clear river, in a hollow, with two fields surrounded by forest and 'ringed trees'" (1894, 165). The practice of ring-barking entails stripping bark from the circumference of a tree trunk, causing its protracted decline and death. Western Australian settlers employed this relatively inexpensive and less labor-intensive technique widely to clear land for pasture and other colonial industries (Western Australia Bureau of Agriculture 1897). An interviewee on the ABC Radio National program *Hindsight* recalls the continuation of ring-barking well into the early twentieth-century in other parts of the state and to clear different eucalypt species: "the tree died and then the branches fell and we spent years and years picking up those branches [...] There were hundreds and hundreds of acres of stark, ring-barked forests" (qtd. in Plate 2005, 266).

It was most likely Mr Brockman or his wife, Capel Carter Brockman, who alerted North to an uncanny karri specimen exhibiting a sizable abnormality known as a *burl* on the lower half of its trunk. Although not specifically mentioning her piqued interest in the deformed tree, her journal (published posthumously as *Recollections of a Happy Life*) does detail the tragicomical circumstances of North's first encounter with the karri tracts west of Pemberton near "the entrance to the forest of perhaps the biggest trees in the world," about eighteen kilometers from The Warren (North 1894, 164). While their carriage was being repaired after careening off the track as a result of the skittish behavior of two ornery horses, the artist had the opportunity to inspect karri trees in a most uninhibited manner: "I spent four delightful hours sketching or resting under those gigantic white pillars, which were far more imposing than the trees of Fernshaw; their stems were thicker and heads rounder than the amygdalina gums" (North 1894, 164). In this passage, North cites the settlement of Fernshaw in the Dandenong Ranges outside of Melbourne, where she observed amygdalina gums growing before coming to Western Australia. Until the publication of Victorian Government botanist Ferdinand von Mueller's *Systematic Census of Australian Plants* (1882) and *Key to the System of Victorian Plants* (1887), the "stupendously tall" (von Mueller 1887, 236) *E. regnans*, or mountain ash (as we know it commonly today) was classified by taxonomists as a variety of *E. amygdalina*, the black peppermint endemic to Tasmania. North characterizes the karris in relation to her observations of other massive eucalpyts occurring in other parts of Australia.

Elsewhere in her journal, North refers to *E. diversicolor* as white gum, with the Nyoongar term *karri* later brought into wider usage in the Southwest region, along with other indigenous tree names, notably jarrah, mallet, marri, tingle, tuart, and wandoo (Abbott 1983, 18). She composes a scene in prose—rather than brushstrokes—of "enormous trees, chiefly white gums, as smooth as satin, and sometimes marbled, with a few rough red trunks or 'black butts' among them, and small casuarinas and shrubby bushes underneath" (1894, 165). From North's optimistic point-of-view, the karri forests—owing to their massive size—would endure the juggernaut of settlement, being "safe to stand because there were no means of taking them away" (1894, 165). Furthermore, notwithstanding the visual gravity of the colossal trees, North maintained a botanist's exactness for particulars, noting "one lovely plant grew quite high, with its leaves arranged in stages all the way up like stars of green, and tiny strings of flowers or waxy berries under each leaf" (1894, 165). Regardless of her recent arrival to the Pemberton area and her grappling with the appropriate words to describe karri country, North exhibits in her journal a combined artistic *and* ecological appreciation of the grand trees that differs significantly to the perception of the forest as appropriable resource or insurmountable obstacle.

An attitude of awe and respect mixed with a touch of hyperbole is evident in "Karri Gums near the Warren River" (Figure 6.1). This rendering of the karri forest is dominated by the burl of a tree at the viewer's right foreground with a grouping of thinner, sleeker karris behind it providing visual contrast to the diseased formation. On the viewer's left, a naturally truncated karri frames the composition with smaller understory trees occupying the middle. A digital version of the painting is included in the Marianne North Online Gallery, hosted by Kew Royal Botanic Gardens, with the smaller trees noted by the gallery curators as casuarinas but which, on close inspection, appear to be banksias. To enhance the scene's dramatic effect, the emus are substantially out of proportion, creating the impression of a monstrous forest that would have astonished European audiences. Having visited the famed Marianne North Tree several times myself,[3] I can attest to this minor painterly solecism. An adult

---

3   Wild Western Australia. "The Marianne North Tree Near Pemberton." *Wild Western Australia*, 2016, accessed December 19, 2017, http://www.westernaustralia-travellersguide.com/marianne -north-tree-pemberton.html.

FIGURE 6.1 *Marianne North. "Karri Gums near the Warren River, West Australia." Early 1880s.*
           Oil on paper. The Marianne North Gallery.
           REPRINTED WITH PERMISSION FROM ROYAL BOTANIC GARDENS, KEW.

forest emu is as tall as an adult man and would be at, or above, the height of the ferns on the ground beneath the burled karri. The painting, nevertheless, embodies the ethereal atmosphere of the forest in the tall, naked, heavenward trunks and wispy crown formations of the karris. The sublimity of the karris contrasts sharply to the grotesqueness of the burl, its oviform heaviness against the vertical levity of the boles. In other words, the burl—a tree growth typically caused by a virus or fungus, in which knots in the grain develop from in-grown buds—brings the composition down to earth in what would otherwise be a quintessential, although slightly exaggerated, karri forest scene.

While North's painting is by all accounts devoid of people, other artists and naturalists sought to include traces of human activity in their forest renderings. Born in Fitzroy, Victoria, in 1853, the naturalist and ornithologist Archibald James Campbell travelled to Western Australia in the late nineteenth- and early-twentieth centuries to conduct pioneering field research on birds, but also produced some of the earliest photographs of karris. Campbell is known as a founding member and former president of the Royal Australian Ornithological Club, as well as editor of their long-running research journal, *The Emu* (McEvey 1979). His Western Australian forays, taken between 1889 to 1920, were published in the Melbourne-based periodical *The Australasian*. For instance, an article from 1890 relates Campbell's observations of the birds, reptiles, marsupials, plants, and Nyoongar knowledge of karri ecology, confirming also that he stayed in the township of Karridale between Augusta and Margaret River during his earliest visit to the region (Campbell 1890). In an article from the 1920s, he exalts the karri the "vegetable giant of the west" (Campbell 1921, 868). However, Campbell is perhaps best-recognized for his generously illustrated, seven-hundred-page tome *Nests and Eggs of Australian Birds* (1901), for which he needed to have travelled to the Southwest of WA for fieldwork, including egg collecting, as indicated in personal observations throughout the text. He relates an anecdote concerning his search for purple loorikeet eggs in the karri forest in October 1889 (1901, 597).

This historical context helps to affirm that Campbell's images of karris are likely some of the earliest taken by naturalists and are some of the oldest extant photographs of karri forests. The images form part of the Campbell collection held by the National Library of Australia, including work between 1870 and 1929. His experimentations in photography and print-making indicate diverse aesthetic sensibilities including not only pictorialized karri trees but a range of species and compositional techniques, some involving other people. Campbell's work demonstrates an intrigue for interactions between plants and animals, even as it lacks an ethics of stewardship or sustainability. In the writings that accompany his ornithological descriptions, for instance, Campbell mentions wantonly killing snakes and other unsavory environmental practices. Nonetheless, his sepia-toned photograph "Karri Forest, Western Australia" presents a quintessential forest scene, vertically composed in an effort to signify the height of the trees, although their tops remain out of purview.[4] The image conveys the challenges of rendering perspective in, and of, the karri forest, both in terms of immediate (direct, sensory, actual) and mediated (compositional, imagistic, textual) modes. The foreground of the image, on first inspection, appears to consist of small zamia palms, or cycads (*Macrozamia* spp.). The karri trunks appear unusually slender, especially those of the background, implying that the scene is a relatively young, regrowth forest. Signs of charring at the lower portions of the karri trunks and the stunted dimensions of the cycads also suggest that fire had razed the forest in recent years. To the surprise of the casual viewer, however, a small, partially concealed figure of a man under the most prominent karri confirms the true scale of the representation and that these are actually old-growth specimens. With his head nearing the height of the surrounding vegetation, including the cycads in question, the man is immersed in the bush with his neck craning upward toward the karri crown out of frame.

Another noteworthy—though questionable—photograph from the Campbell archive also attempts to capture the soaring height of what is claimed to be a karri with uncharacteristic multiple lower trunks, perhaps being overtaken by a strangler tree of unknown identity. Campbell's sepia-toned "Karri Tree,"[5] from either the late 1800s or early 1900s, derives its title from the pencilled inscription on the reverse of the original. However, the naming of the species might be inaccurate. Despite the zamia cycads of the foreground, the central tree and surrounding forest appear closer in composition to beech myrtle (*Nothofagus cunninghamii*) communities found in Tasmania and Victoria. In my eight years of regularly visiting karri country, I never observed such crowded entanglements of vegetation in what is ordinarily the clear and clean space of the karri understory. The photo's arrangement does point to one of the principal technical challenges negotiated by early photographers of Australia's tall eucalypt forests: how to encompass the full extent of the tree within the visual constraints of the camera lens. Some early photographic depictions of karris endeavor to resolve the issue by splicing three or four individual images together to encompass a tree from top to bottom in a single

4   National Library of Australia. "Karri Forest, Western Australia [2]." *Trove*, 2017, accessed December 19, 2017, http://nla.gov.au/nla.obj-147133725/view.
5   National Library of Australia. 'Karri Tree.' *Trove*, 2017, accessed December 19, 2017, http:// trove.nla.gov.au/version/29886277.

vertical sweep. This is true of historic photographs on display at the Pemberton Visitors' Centre in 2016: a rendering of the Gloucester Tree with its telltale bushfire lookout platform; an elongated image of a huge karri that depicts the complete vertical extent of the tree in order to accentuate the sawman's prowess; and a photograph of a well-dressed, upper class party posing at the base of a karri with its entire bole and crown synoptically composed.

Campbell's "Karri Tree, 254 Feet in Height, Western Australia"[6] features two men poised at the base of a gigantic specimen. In keeping with his other images of karris, the human figures add proportion to the composition by communicating the tree's actual enormity. The taller and more darkly dressed of the two individuals is positioned sideways to the camera, while the other turns—his weight shifted to his left leg—to face the photographer, presumably Campbell. The second figure grips a measuring tape wrapped at least partially around the mammoth's circumference. The men's empirical interest in the karri is motivated not by a naturalist's inquisitiveness or an artist's sense of awe, but rather by the prospect of converting the tree into profitable timber—and thus the physical challenge that lies ahead for them. The ambivalent semantics of the photograph parallels the ambivalence of settlers towards the karris as simultaneously objects of admiration and conquest (see chapters 1, 2, and 3). Images, such as this, also embody the Enlightenment-based impulse to quantify the natural world in order to exert control over an otherwise unknown, unruly, and potentially threatening and alienating non-human domain (Gascoigne 2002).

Not all of Campbell's renderings of the karri forest relate practices of measurement, quantification, progress, and, ultimately, conquest. The Campbell archive also contains delicate, skillfully produced prints that demonstrate an innate curiosity for the karri environment. Attributed to Campbell, "In the Karri Forest"[7] is a black and white print exhibiting a diverse range of textures—from the vertical patterns of the karri boles to the horizontal and transverse strokes of the forest foliage. The prominent absence of human impacts and figures—Nyoongar or Anglo-Australian—invokes the wilderness tradition in environmental representation, later instantiated in the landscape photography of Tasmanians Olegas Truchanas (1923–72) and Peter Dombrovskis (1945–96) (Giblett and Tolonen 2012, 93–102). Another print, "Nest of the Grey-breasted

---

6   National Library of Australia. "Karri Tree, 264 Feet in Height, Western Australia." *Trove*, 2017, accessed December 19, 2017, http://trove.nla.gov.au/version/29846589.
7   National Library of Australia. "In the Karri Forest." *Trove*, 2017, accessed December 19, 2017, http://trove.nla.gov.au/version/34086000.

Robin, a Karri Giant,"[8] juxtaposes two complementary visual perspectives on the forest. The first, on the viewer's left, is an up-close rendering of grey-breasted robin's (*Eopsaltria griseogularis*) nest, protected in the crook of a balga tree (*Xanthorrhoea preissii*). The viewer's right panel takes the perspective of "Karri Tree, 254 Feet in Height," in showing two men sizing up a large karri with its characteristically variegated lower bark pattern. The karri dwarves the two men as they go about indeterminable business. The viewer oscillates between an intimate sense of non-human dwelling—or *oikos*—through the close rendering of the nest and, in comparison, a sublime, distanced construction of space in the depiction of the karri forest and the undisclosed surveying activities of the male figures.

Early Twentieth Century (1901–45): Federation to World War II

The early twentieth century ushered in heightened visual focus on karris, in some measure due to advances in image-making technologies. This was combined with the intensification of economic activities in the forests and a sustained interest in karri timber as an international export. The works of this period also begin to exhibit traces of car-based, pleasure tourism in karri forests. Representative of the environmental context of the early 1900s is "Karri Forest," taken between 1900 and 1909 by an unidentified photographer. Part of the National Library of Australia's Historical Records Rescue Consortium (HRRC) archival initiative, the work consists of two glass negatives with the first portraying a relatively undisturbed eucalypt habitat, deficient of any sign of human presence and organized pictorially, as in some of Campbell's images, around an especially large karri centerpiece specimen.[9] The second negative departs markedly from the first's sylvan idyll by pictorializing the consequences of logging. The image shows a conspicuous stump in the foreground, a partly felled tree resting obliquely at the image's left side and the sparser background of a forest that has borne—and is in the process of bearing—intensive human impacts. Another black and white negative on glass, "Karri Falling" (1910),[10] also exemplifies the imagery of karris produced during this period: tree fellers

---

8   National Library of Australia. "Nest of the Grey-breasted Robin, a Karri Giant." *Trove*, 2017, accessed December 17, 2017, http://trove.nla.gov.au/version/34234230.

9   National Library of Australia. 'Karri Forest.' *Trove*, 2017, accessed December 17, 2017, http://trove.nla.gov.au/version/174383503.

10   State Library of Western Australia. 'Karri Falling.' *Trove*, 2017, accessed December 17, 2017, http://trove.nla.gov.au/work/14515140?q=karri&c=picture.

flee the scene of a giant karri beginning to collapse. Some of these themes, and others, play out further in the works of the painter Henry Prinsep (1844–1922), print-maker Henri van Raalte (1881–1929), and the pioneering photographers Axel Poignant (1906–86) and Frank Hurley (1885–1962).

A colonial civil servant, estate manager, horse-trader, artist, and photographer born in India, Henry Charles Prinsep travelled extensively around the Southwest region of Western Australia in his younger years with a view to profit from jarrah and karri exports for the development of the Indian railway system (Allbrook 2014, 134). His early land-based capitalistic exploits, however, were not entirely met with success. For instance, in 1870, the ship *Hiemdahl*, loaded with Prinsep's jarrah sleepers and horses bound for India, wrecked in the Hooghly district of West Bengal, plunging him into several years of financial disarray (Staples 1988). He later worked in a variety of administrative positions, including an appointment as chief protector in the new department of Native Affairs in 1898. Within his economic and bureaucratic ventures, he maintained an ongoing interest in art and served as founding member of the West Australian Society of Artists. During his retirement at Busselton, WA, Prinsep constructed his own darkroom and continued producing art of the Southwest landscape, including the trees he formerly viewed in utilitarian terms. Prinsep's paintings, prints, and photography often show Anglo-Australians dwarfed by colossal karris. This disproportion between karris and humans signifies the tree's value as a viable economic resource while documenting the consuming effects of settler expansion into karri country (Allbrook 2014, 205).

One of Prinsep's most recognized works is the oil-on-canvas painting "Karri Trees, Manjimup c.1910"[11] held in the Wordsworth Collection at the National Gallery of Australia (see Figure 4.1). Completed later in Prinsep's life, the piece represents a newer style for him. In comparison to his earlier sketches and renderings, "Karri Trees" presents a bold, modernist Federation-era image of the stout forms of Southwest trees—particularly the central karri—counterbalanced by the acute horizontality of the cleared landscape behind the trees and the fence of the mid-ground.[12] Unlike North's late-nineteenth-century forest scene of emus, banksias, and karris, free from traces of colonial impacts, Prinsep's painting brings to prominence human intercalations by juxtaposing the crowded intermingling of karri country to the relative sparsity of the settler homestead. In contrast to forest scenes dominated by the

---

11   "Karri Trees, Manjimup c.1910" in Allbrook (chapter 10). Page down for the image in full: http://press.anu.edu.au/apps/bookworm/view/Henry+Prinsep%E2%80%99s+Empire%3A+Framing+a+distant+colony/11161/Text/ch10.xhtml.

12   NLA commentary.

strong up-and-down patterns of karris, Prinsep's rendering highlights the intersections: verticality and horizontality, entanglement and disentanglement,
environment and settlement, nature and culture. Yet, rather than concede to
colonization and its impacts on country, the painting foregrounds the dynamic
intermixing of forest species that co-exist with karri trees, including zamia cycads, ferns, vines, and other herbaceous plants. The visual message is perhaps
one of resilience and hope, suggesting that despite the intensity of logging and
other extractive activities in the region, karri ecology will endure in the spaces
between human and non-human communities (for an alternate reading of the
painting, see chapter 4).

The pioneering Australian etcher and print-maker, Henri van Raalte, also
chose *E. diversicolor* as a subject in his "A Karri Tree Glade," created between
1917 and 1919 (Figure 6.2). Born in London and migrating to Western Australia
in 1910, van Raalte initially took paid work as a timber-getter, during which he
was exposed to the diversity of eucalypt species in the Southwest landscape.
These early experiences of the region's bush imprinted distinctively upon his
creative work, much of which reinterpreted old print-making techniques in the
new context of the Australian milieu. As an artist keenly attracted to Western
Australian gums, specifically karris and tuarts, van Raalte prioritized imagina-
tive figurations of trees, rather than humans, in his etchings. Thus, van Raalte's
work, on the whole, counterpoises the preponderance of photographic imagery during this period in which human figures—although obviously smaller
in stature—nevertheless dominate the hulking karris with saws and other
mechanical extensions of colonial advance. His first major gum tree etching,
"The Monarch on the Kalga [sic] River, WA"[13] from 1918, epitomizes the mood
and style characteristic of his work. The print presents an enchanted scene of
a tuart (*E. gomphocephala*) along the Kalgan River in the Great Southern region of Western Australia between the towns of Denmark and Albany. Another
print, "Tuarts of the South-West" (1920), is a Tolkienesque scene done in brown
ink and featuring the endemic tuarts of the region.[14] Van Raalte also practiced
aquatint (an intaglio print-making technique involving a copper or zinc plate)
and drypoint (also an intaglio technique in which an image is imprinted onto
a plate with a metal- or diamond-pointed needle). Considered a seminal figure
in Australian etching, van Raalte facilitated the development of print-making

13   Art Gallery of New South Wales. "'The Monarch' on the Kalga [sic] River, W.A." *Art Gallery
        NSW Collection*, accessed December 17, 2017, http://www.artgallery.nsw.gov.au/collection/
        works/6314/

14   National Gallery of Australia. "Tuarts of the South-West." *Artsearch*, accessed December
        17, 2017, http://artsearch.nga.gov.au/Detail-LRG.cfm?IRN=39138&PICTAUS=TRUE.

FIGURE 6.2 *Henri van Raalte. "A Karri Tree Glade." 1917–1919. Aquatint and Drypoint*
printed in brown ink. 40.1 × 25.4 cm (plate) 51.1 × 35.0 cm (sheet). Spartalis 40.
ed. 2/60 National Gallery of Victoria, Melbourne presented by the Estate of
Mrs Cutten, 1992 (P60–1992).

in the inter-war period and later became one of the first artists to develop a process for colour etchings (Chappell 2016). "A Karri Tree Glade," an aquatint and drypoint in brown ink, evokes the preternatural style of his tuart prints. The work poignantly envisions the enchantment of a karri forest—its elusive, mysterious essence, at dusk or dawn—as a stark departure from the realism of North and Prinsep, or the classificatory approach to nature evident in Campbell's photography.

British-born photographer Axel Poignant migrated to Australia in 1925 under the Dreadnought scheme, which brought young men from Britain to work on farms in New South Wales. After completing itinerant jobs in Sydney and outlying parts of the state, he made a name for himself as a photographic portraitist and later contributed, with Stuart Gore (1905–84), one of Western Australia's first aerial photographers, to the production of an aerial survey of the Goldfields region of Western Australia (Sassoon 2012). He also became well-known for his perceptive photographs of Australian Aboriginal people, particularly of Arnhem Land. Through the encouragement of the naturalist Vincent Serventy in the 1930s, Poignant extended his photographic reach from human portraits to Australian natural history. Much of his work came to address the interconnections between people and land. His celebrated photograph "Australian Swagman"[15] of an elderly man with his back to the camera pushing a bicycle loaded with his possessions along a deserted road demonstrates Poignant's identification with the tenacity and self-reliance of rural people.

A master of portraiture and narrative, Poignant created a body of work recognixed for its style and selection of angles (Sassoon 2012). This sophistication holds true for his images of karri forests from the 1930s. Rather than constructing an idealized, "untouched" forest scene centering on a single grand gum specimen, the photograph "Logging in the Karri Forest, Pemberton, Western Australia, September 1935"[16] attends to forestry activities, highlighting the interactions between a group of male timber-getters and the karri landscape. The drama centers on the diminutive figure of a man ascending—probably not descending, if other images in the narrative are any indication—a karri soon-to-be felled, recognizable as such for the climbing spikes spiralling up

15   National Gallery of Australia. "Australian Swagman." *NGA Federation Australian Art and Society 1901 to 2001*, 2017, accessed December 17, 2017, http://nga.gov.au/federation/Detail .cfm?WorkID=34603.

16   National Library of Australia. "Logging in the Karri Forest, Pemberton, Western Australia, September 1935." *Trove*, 2017, accessed December 17, 2017, http://nla.gov.au/nla .obj-147825658/view.

the tree's bare trunk. Other workers appear at ground level, inspecting downed logs and posing as if at momentary rest from the ongoing labor of clearing—or as if preparing for the imminent collapse of the tall gum. At the bottom left corner of the composition, a tin-roofed hut is a hive of activity on the bustling forest floor. The portrait is less about the forest ecosystem—as in Campbell's photography—and more about human economic preoccupations. Indicative of Poignant's style, the image furnishes a fixed instance from a larger narrative. "Axeman Cutting the Top of the Tree in the Karri Forest, Pemberton, Western Australia" relates to the previous image's story line in its framing of a logger, having completed his ascent, extending his axe back in preparation for felling the tree top—all set against the sky's brooding darkness.[17] A third image, "The Cut Top of the Tree Falls in the Karri Forest, Pemberton, Western Australia, ca. 1934,"[18] completes the sequence by portraying the toppling of the karri crown, with the climbing man's silhouette hunkered down in a safe position below the main breaking point. The tree is rendered as a colorless, solid black adumbration, in contrast to other depictions of karris that accentuate the textural variations of the bark.

While Prinsep and van Raalte were by all accounts Western Australians, having lived the better part of their lives in the state, and Poignant had a clear Australian affinity, other artists with affections for karris were more global in their creative pursuits, as in the tradition of Marianne North. This is especially true of Frank Hurley, who has been called by critics "Australia's most celebrated early landscape photographer" (McDonald qtd. in Giblett and Tolonen 2012, 83). Born in Sydney where he developed a reputation for his technical knowledge and photographic risk-taking, Hurley was appointed in 1911 by Sir Douglas Mawson as the official photographer of the Australasian Antarctic Expedition, and later served in the Australian Imperial Force during World War I. His landmark photographic works, such as "Morning at Passchendaele,"[19] exhibited the development of image-making techniques as well as his disdain for the war (Pike 2016). He went on to take photos and make documentary films in Papua, the Middle East, and other parts of the globe, and, in 1953, he published

---

17   National Library of Australia. "Axeman Cutting the Top of the Tree in the Karri Forest, Pemberton, Western Australia." *Trove*, 2017, accessed December 17, 2017, http://nla.gov.au/nla.obj-147829760/view.

18   National Library of Australia. "The Cut Top of the Tree Falls in the Karri Forest, Pemberton, Western Australia, ca. 1934." *Trove*, 2017, accessed December 17, 2017, http://nla.gov.au/nla.obj-147829865/view.

19   National Library of Australia. "The Morning After the First Battle of Passchendaele." *NLA Digital Archive*, 2017, accessed December 17, 2017, http://nla.gov.au/nla.pic-an24574133.

Western Australia: A Camera Study (Hurley 1953), containing some karri imagery. It is during this later, more settled, time of life that Hurley turned his eye toward the tall forests of the Southwest, including *E. diversicolor*. "Giant Karri Trees in the Diamond Tree Forest,"[20] most likely taken in the 1940s or '50s during the production of his photobook *Western Australia*, is part of the vast Hurley negative collection. The image bears the name of the Diamond Tree, a gargantuan karri located ten kilometers south of Manjimup. The tree is well-known for its use as a fire lookout from 1941 to 1973, and is now is one of three lookout trees popular with tourists who venture to the top. The image's vertical composition echoes Prinsep's "Karri Trees, Manjimup c.1910" in its contrast between the bold figures of the karris of the foreground and the empty horizontality of the cleared space of the background.

During this period, images of karris increasingly depict encounters between human and ecological communities. Hurley's photography evokes these intersections, particularly as they manifested as networks of former logging tracks turned tourist routes through forested country. The image "Karri Forest, Pemberton [Western Australia, 2]"[21] documents the beginnings of the Southwest botanical tourism industry during the mid-twentieth century. A conspicuously large karri frames the left side as an automobile rounds the corner of a well-groomed road—possibly a precursor of contemporary tourist thoroughfares, such as the Karri Forest Explorer[22] and others, allowing public access to karri country with much more comfort than in Marianne North's days. The image captures the transition from extractive to tourism industries in the Southwest, heralding the formation of an extensive system of parks and reserves that continues to draw visitors from Australia and abroad. In a similar vein but in stylistic contrast to depictions such as van Raalte's, "Karri Forest, Pemberton [Western Australia, 3]" centralizes the figure of a woman with her back to the camera standing in the middle of the road as a car turns the corner toward her.[23] A roadside post, perhaps a mile-marker, with a white-painted top is situated directly behind the standing woman and signifies the complexities of "progress" and development in the forests. The karris of this landscape are

20   National Library of Australia. "Giant Karri Trees in the Diamond Tree Forest." *Trove*, 2017, accessed December 17, 2017, http://trove.nla.gov.au/version/40098416.

21   National Library of Australia. "Karri Forest, Pemberton." *Trove*, 2017, accessed December 17, 2017, http://nla.gov.au/nla.obj-157591036/view.

22   Department of Environment and Conservation. "Karri Forest Explorer." *Department of Parks and Wildlife*, 2013, accessed December 17, 2017, https://parks.dpaw.wa.gov.au/sites/default/files/downloads/parks/20120463karriforestexplorer_web.pdf.

23   National Library of Australia. "Karri Forest, Pemberton [Western Australia, 3]." *Trove*, 2017, accessed December 17, 2017, http://nla.gov.au/nla.obj-157592476/view.

considerably thinner than those of "[Western Australia 2]" and thereby permit the late afternoon sun to illuminate the white boles, creating a distinct, quint-essentially Western Australian, lighting effect.

Late Twentieth Century (1945–88): World War II to Bicentenary

The second half of the twentieth century brought technical innovations in color photographic imagery that made possible new visual interpretations of karri forests. Cameras became less the domain of specialists and, subsequent-ly, more accessible to the public, as seen by the increased uptake of image-making technology by writers, travellers, and naturalists. Artistic advances in modernism and abstraction began to inform depictions of karris. As styles, genres, techniques, and media diversified, so too did the art of karris, allowing innovative forms of representation to influence the public imagination. This section will cover the period approximately spanning the beginning of World War II to the Australian Bicentenary in 1988. The photography of Western Australian novelist John Ewers (1904–78) and botanist John Beard (1916–2011), as well as the paintings and prints of artist Guy Grey-Smith (1916–81) demon-strate key themes and developments during this time.

Born in the Perth suburb of Subiaco, the novelist, poet, and critic John Keith Ewers became prominent in Perth literary circles during the World War II era and later served as founding member of the Western Australian branch of the Fellowship of Australian Writers. His novels include *Money Street* (1933), *Tales from the Dead Heart* (1944), and *Men Against the Earth* (1946), the latter being an historical narrative with ecological themes originally forming part of a larg-er manuscript about his experiences as a teacher in the Wheatbelt in the 1930s. Ewers also published short non-fiction travelogues in *Walkabout*—a forerun-ner of *Australian Geographic* existing between 1934 and 1974 that integrated natural history and social observations with travel writing as an outlet for early Australian photojournalism. In an article in *Walkabout*, Ewers gives an account of the stocking of trout in Big Brook, near Pemberton, while also characterizing the generous water systems of karri country: "There was an abundance of per-manent running water, each tiny gully sending its contribution of spring-fed streams into the brooks and rivers that threaded their way among giant karri" (Ewers 1946, 35). His photos of pleasant recreational scenes in karri country, as in "Early Morning at the Swimming Pool Seen from Among the Karri,"[24]

---

24   National Library of Australia. "Early Morning at the Swimming Pool Seen from Among the
      Karri." *Trove*, 2017, accessed December 17, 2017, http://trove.nla.gov.au/work/192056877?q
      &versionId=209922681.

FIGURE 6.3 *John Ewers. "Karri Logs in the Forest, Pemberton, 1950s." Sourced from the Collections of the State Library of Western Australia and reproduced with the permission of the Library Board of Western Australia.*
IMAGE NUMBER B2908293_001.

provide a welcome counterpoint to the profusion of land clearing images during this period.

Most of Ewers' visual works available in the State Library of Western Australia archives are color photographs portraying karri logging in detail. He presumably shot "Karri Logs in the Forest, Pemberton, 1950s" during a fact- and observation-gathering trip to tall forest country in preparation for travelogue writing (Figure 6.3). Reflecting themes of his novel *Men Against the Earth*, the image—rather than naively glorifying forestry practices or attempting to document the process objectively—hones in narratively on the disjunctures between economic interests, local livelihood, and old-growth ecosystems. The efficiently cut logs appear in orderly rows with their sawn ends positioned toward the camera, as a standing karri frames the left side of the image. The lone figure of a worker, with only his back in view, inspects the scene and—apparently lost in some kind of contemplation—shifts his weight to his right leg and braces his left hand against one of the fallen karris. A small equipment shed, the trace of an access road, and the green thickness of the forest are visible in the background.

The composition of Ewers' images hinges on the polarizing juxtaposition between the vertical orientation of the forest against the "unnatural" horizontal dominance of the fallen logs and barren ground. This visual intersection is further reflected in "Tractor Hauling Karri Log, Pemberton, W.A., 1950s,"[25] a four-photo, narrative sequence that documents advances in motorized logging equipment represented by a tractor and its insidious steel grapple—the jaws of death—extension used to clasp and drag logs. The series' first image offers a ground-level view of a thoroughly debarked tree being dragged behind a tractor—itself out of the frame—along a gritty road. A dirt cloud created by the operation obscures a vehicle parked in the bush off to the right. The process unfolds additionally in the second photograph as the tractor hauls the log between two standing trees alongside the track. The dusty residues persist. Perspective shifts in the third image, a front-end view of the tractor including the darkened image of its driver, with the desolation of the surrounding landscape appearing more pointedly reminiscent of a warzone or disaster management area. The concluding snapshot furnishes a close-up view of the log-hauling mechanism as the tractor begins straining up an incline. Ewers' photography exhibits a novelist's sensitivity to narrative combined with the advances in color production and exposure time that arose during the post-World War II period of Australia.

Botanist John Beard also made use of emerging photographic technologies, as evident in his sizeable collection of karri images taken in support of his scientific field studies of Western Australian flora. Born in London and educated at Oxford, Beard became the Director of Kings Park and Botanic Garden in Perth during the 1960s and published throughout his long career influential scientific papers and popular texts. Some of his *E. diversicolor* captures are reproduced in the landmark *Plant Life of Western Australia* (Beard 1990), containing over five-hundred plant photos from various locales around the state. "Standing on Felled Karri Tree, Western Australia," dated between 1965 and 1984, depicts Beard with his back to the camera gazing down at the freshly cut, reddish-tinged karri stump on which he stands (Figure 6.4). The sawdust piles adjacent to the tree base suggest that the toppling of the giant occurred just momentarily before the photo. A chainsaw rests on the stump while a worker in the background raises his axe in preparation for further processing the downed log. As he casts his eyes over the tree's exposed growth-rings, signifying the vast age of the specimen, Beard's demeanor is distinctly

---

25   State Library of Western Australia. "Tractor Hauling Karri Log, Pemberton, W.A., 1950s." *SLWA Online Catalogue*, 2017, accessed December 17, 2017, http://purl.slwa.wa.gov.au/ slwa_b2908293_002.

FIGURE 6.4 *John Beard. "John Beard Standing on a Felled Karri tree, Western Australia,*
1965–1984." Sourced from the Collections of the State Library of Western Australia
and reproduced with the permission of the Library Board of Western Australia.
IMAGE NUMBER B4173080_1.

contemplative, nearly forlorn, and slightly melancholic. He also, however, pho-
tographed more idyllic and familiar karri country scenes, as in "Lookout on the
Gloucester Tree, Pemberton, Western Australia"[26] and "Tall Forest of Karri,"[27]
both appearing in his *Plant Life of Western Australia.*

Born at East Wagin, WA, Guy Grey-Smith became one of the most significant Australian painters, print-makers, and ceramicists of the twentieth
century and a pioneer of modernism in the visual arts in Western Australia
after a period spent studying with famed sculptor Henry Moore in London
(Gaynor 2012). Following a thirty-five year career that geographically spanned
Australia, England, and Southeast Asia, Grey-Smith died in 1981 in Pemberton,

---

26   National Library of Australia. "Lookout on the Gloucester Tree, Pemberton,
      Western Australia." *Trove,* 2017, accessed December 17, 2017, http://trove.nla.gov.au/
      version/211677922.

27   National Library of Australia. "Tall Forest of Karri." *Trove,* 2017, accessed December 17,
      2017, http://trove.nla.gov.au/work/192944480?q=john+beard+karri&c=picture&version
      Id=211212415.

after having retreated to the Southwest region in the mid-1970s with his wife, artist Helen Grey-Smith, to escape the advance of suburbanization near his long-time home in Darlington in the Perth Hills (Gaynor 2012, 1). With his first *E. diversicolor* artworks dating back at least to 1951 with the oil-on-canvas "Karri Forest" (Crawford and Crawford 2003, iv) and to 1958 with "Karri Forest" (Gaynor 2012, 202), he pursued karri trees as artistic subjects throughout his career. Grey-Smith's life-long attention to the tall gums of the Southwest parallels his interpretation of modernist techniques and styles, executed in a range of media, for evoking the visual characteristics of karri country. The media of the following examples of Grey-Smith's karri work vary between oil and beeswax emulsion, serigraphy (also known as screen printing, silk-screening and serigraph printing in which ink is applied to a substrate using mesh), woodblock printing (or woodcut printmaking involving relief printing techniques), and linocut (using a linoleum sheet, sometimes on a wood block, as the relief surface). While the media are innovative, many of the works' titles are generic variations of the phrase "Karri Forest."

Dated from 1956, "Karri Forest"[28] is an oil on canvas painting measuring sixty-four by forty-nine centimetres in portrait orientation. This early example of Grey-Smith's karri enthrallment presents a less abstracted rendering in which the normal, perceptual features of the environment are evident in the lofty tree crowns, tall naked trunks and forest floor vegetation. However, twenty years later, we find a marked evolution of Grey-Smith's style in the 1976 painting "Karri Forest,"[29] an oil and beeswax emulsion applied on gauze over hardboard of about sixty by ninety centimeters (thirty-five inches). An abstract, geometrical scene embodying the influence of Post-Impressionists such as Cézanne, the bold up-and-down strokes—signifying the karri boles—reflect the compositional tendencies of previous eras and traditions. The painting departs sharply in other respects from what we have seen so far in this chapter in its rough textures and fissuring of the dried paint mimicking karri, jarrah and marri bark. Grey-Smith also employs relatively homogenous fields of blue, red and green against which thin red, yellow and light blue streaks run vertically, interpreting, through a modernist mode, the play of light in karri forests. "Karri

---

28   Arcadia Auctions. "Some Works of Guy Edward Grey-Smith." Arcadia Auctions Online, 2015, accessed December 17, 2017, http://www.arcadja.com/auctions/en/grey_smith_guy_edward/artist/45810/

29   Art Gallery of Western Australia. "Karri Forest." *AGWA Collection Online*, 2017, accessed December 17, 2017, http://www.artgallery.wa.gov.au/exhibitions/guy-grey-smith.asp.

Forest II,"[30] most likely created during the 1970s, appears also to be oil and beeswax rendering that implies the clearing of the forest in the visual dominance of a blue field framed by elongated brown pillars over an empty and flat green surface.

Grey-Smith also applied modernist abstraction to his karri wood- and screenprints. Produced in 1975 in Darlington just before he moved to Pemberton, "Karri Forest Foliage"[31] is a linocut from one block printed in black ink. The dome-shaped leaf abstractions thread organically on an oblique axis throughout the frame. "Karri Forest"[32] from 1980 is a single-block woodcut printed in black ink. Rather than the diagonally-composed foliage filigree of the previous work, "Karri Forest" employs bold vertical patterns to emphasize the solidity of the karri forest and its upward orientation. The effect is one of Wagnerian ascendancy countering the scenes of desolation dominating karri photography during the mid-twentieth century. Finally, "Karri Forest" is a color screenprint from 1973, in shades of purple, yellow, blue and red, that balances the karri forest compositionally along x- and y-axes.[33] The lower half of the print consists of upright patterns of tree figures whereas the upper portion comprises the circular swoops of the canopy topped by an expanse of blue sky. Grey-Smith's experimentation with different styles and media exemplifies the significance of karri trees to Western Australian artists throughout the twentieth-century and into the twenty-first.

Contemporary Period (1988–Present): Bicentenary to the Present

In the years since the Australian Bicentenary, karris continue to resonate as prominent subjects and symbols for artists working in a variety of media, including painting, photography, and print-making. In the broad survey of artworks presented thus far in this chapter, the significance of karris to the trajectory of Western Australian art is striking. Karris have been central the creative imagination of artists and the public over time. Although the

---

30   ABC. "Karri Forest II." *Radio National Online*, 2017, accessed December, 17, 2017, http://www.abc.net.au/radionational/programs/booksandarts/karri-forest-ii/4305560.

31   National Library of Australia. "Karri Forest Foliage." *Trove*, 2017, accessed December 17, 2017, http://trove.nla.gov.au/work/14515148.

32   National Library of Australia. "Karri Forest." *Trove*, 2017, accessed December 17, 2017, http://trove.nla.gov.au/version/17117666.

33   Guy Grey-Smith. "Karri Forest." *Invaluable*, 2015. Accessed December 17, 2017, https://www.invaluable.com/auction-lot/guy-grey-smith-1916-1981-karri-forrest-screen-52-c-bf84768b52.

techniques, styles, and messages have changed, karri art thrives in the present, as the works of Howard Taylor (1918–2001), John Austin (b. circa 1950), Tony Windberg (b. 1966), H.E. Quicke (b. circa 1930), and Carol Thompson will attest. Although other well-known contemporary Australian artists, such as John Olsen (b. 1928), have engaged with *E. diversicolor*, the works of Taylor, Austin, Windberg, Quicke and Thompson are indicative of current trends.

A greater concern for the natural world in karri art in recent years reflects the advent of the environmental movement in Western countries during the late 1960s and '70s. A shift in values emphasis from economics and utilitarianism to ecology and sustainability within parts of the general population parallels concerns of conservation, natural history, and connection to place. As evident in Austin's photography, karri art has begun to incorporate activist themes emanating from the old-growth forest protection campaigns of the 1990s, 2000s, and present decade. Alongside the rise of ecological messages within the art of karri country, there has been a renaissance of tourism in the Southwest, propelling a nature art industry that takes karris as one of its preferred subjects. Whereas traditional landscape painting fell somewhat out of favor by the mid-twentieth century with the ascent of modernism in the works of Grey-Smith and others, the region's vibrant tourism industry has contributed to renewed focus on realistic depictions of karri communities for the souvenir and fine-art-collecting markets. A third trend of note in contemporary karri art, especially in Windberg's paintings, is the use of forest substances, such as charcoal and ash, as media, resulting in a powerful sense of place within the very materiality of the representations.

Born in Hamilton, Victoria in 1918 and relocating to Perth in 1932, Howard Taylor would achieve recognition as one of the most influential Australian painters, potters, sculptors, and graphic artists of the twentieth century (Snell 1995). After being held as a prisoner of war in Europe during service with the Royal Air Force, Taylor returned Bickley, in the Perth Hills, then moved to Northcliffe in 1967 where he remained for the rest of his life. His artistic practice was characterized by slow, patient observation of natural phenomena—a technique he perfected in the reclusion of Northcliffe.[34] Like Guy Grey-Smith, experimentation and innovation were the hallmarks of his approach to the Western Australian bush. During the 1980s and '90s, Taylor intensified themes of light and space, forest figures, landscape colonnades, natural phenomena, and forest regeneration—all of which manifest on a daily basis in the

34   National Portrait Gallery. "Howard Taylor." *National Portrait Gallery Online*, 2017, accessed December 17, 2017, http://www.portrait.gov.au/portraits/2001.175/howard-taylor.

remoteness of karri country.[35] As critic Christopher Heathcote comments, Taylor "pared his experience of life in a Karri forest to fleeting white-on-white abstractions" (2002, 22).

"Karri Forest 1962"[36] is an oil-on-board painting that represents one of Taylor's earliest renderings of the forests. The pronounced earth tones— browns, tans, olives—mix with streaks of white and black to evoke a tall euca-lyptus scene with an open area in the foreground suggesting the presence of a road or clearing. In its modernist underpinnings, the bold, thin, vertical strokes are reminiscent of Grey-Smith's abstracted karri woodprints. Similarly, "Karri Forest Painting"[37] from 1965 is an oil and acrylic on wood panel that, despite its abstracted forms, retains semblances and hints of karri trees in the marked verticality of the brushstrokes. Later works, however, such as the 1993 oil paint-ing "Tree Line with Green Paddock" are in all likelihood interpretations of karri country but lack any explicit references to the trees.[38] The painting consists of three horizontal bands: the top bears rough semblances of trees; the middle is a homogenous yellow-green zone; and the bottom has a similar pattern and tone as the top, but without the sylvan figures. Likewise, the oil-on-canvas "Study for Forest Land"[39] constitutes an investigation of forest light effects but altogether avoids literal references to karris. Instead, the movement between darkness and illumination, between spaces of shadow and sun—which is characteristic of human perceptual experience of the karri forest—informs the painting.

Photographer John Austin has been taking black and white images of the Australian landscape since the 1970s. He came to settle in the township of Quinninup, near Pemberton, where he documented artist Howard Taylor at work in his studio, and also became involved as a photographer in the activist campaigns to protect old-growth tracts from logging. Whereas much of Austin's work documents the destructive impacts of logging alongside public efforts

35   Howard Taylor, Phenomena. "Phenomena Exhibition." *Phenomena Online*, 2017, accessed December 17, 2017, http://artgallery.wa.gov.au/exhibitions/howardtaylor/exhibition.asp.

36   Australian Art Sales Digest. "Karri Forest 1962." *Australian Art Sales Digest Online*, 2017, accessed December 17, 2017, http://www.aasd.com.au/index.cfm/list-all-works/?concat= taylorhowar&order=1&start=101&show=50.

37   Galerie Dusseldorf. "Karri Forest Painting." *Galerie Dusseldorf Online*, 2017, accessed December 17, 2017, http://www.galeriedusseldorf.com.au/GDArtists/Taylor/HT60sExh/ HT60sMore/HTTRCanopy.JPG.

38   Deutscher and Hackett. "Tree Line with Green Paddock." *Deutscher and Hackett Gallery*, 2017, accessed December 17, 2017, http://www.deutscherandhackett.com/ auction/40-important-australian-international-fine-art/lot/tree-line-green-paddock-1993.

39   Galerie Dusseldorf. "Study for Forest Land." *Galerie Dusseldorf Online*, 2017, accessed December 17, 2017, http://www.galeriedusseldorf.com.au/GDArtists/Taylor/HTExh2002/ HHTExhGD2002Images/HHT0045W.JPG.

to protect karri country, other images offer more contemplative reflections on the intrinsic beauty and spirit of the forest. In particular, three photographic portfolios from Austin's website exemplify a new ethos in karri representation evident in contemporary times—one concerned with matters of forest preservation beyond the aesthetics, utilitarianism, economics, spiritual aspects or the symbolic dimensions of the trees. These portfolios are "Australian Forest: Karri, Marri and Jarrah" (1983–2010),[40] "Karri Forest Logging" (1994–2010),[41] or what he also calls "Forest Threnody" denoting a song of lamentation, and "Forest Protest and Activism" (1994–2003).[42]

The first portfolio, "Australian Forest," presents an array of perspectives on the forests at various times of day, including in the dawn mist, as well as close-range images of the diverse species and ecosystems that comprise karri country: zamia cycads, balgas and, notably, a melaleuca swamp comprising the bulk of one black and white photograph. Rather than representing karris as sublime objects in visual isolation from their broader habitats and companion species, Austin conveys an ecological sensitivity to the treescapes as communities in the sky and on the ground. Indeed, the series features few signs of invasive human impacts. However, by the final image in the portfolio, a caption reveals the back-story. Boorara Forest, the site of several karri and marri photographs imparting the impression of pastoral harmony, was in fact clearfelled in 2000 despite heavy opposition by local environmental groups. It turns out that Austin's photographs are among the last taken of the place, which remains solely in memory, imagination and imagery (see chapter 8 for a history of forest activism in Southwest WA old-growth country). "Forest Threnody" continues the sharp departure from the idyll of the first series by foregrounding the assault waged against the karris by the logging industry. Its opening image, entitled "Peta and Debbie on the Megastump, Gardner 08 State Forest, January 1999," features the naked bodies of two protestors curled in foetal positions on an enormous tree stump measuring four meters in diameter and fourteen in circumference.[43] According to Austin's caption, the timber of the immense karri was later deemed useless and discarded on the forest floor. "Forest Protest

40   John Austin. "Australian Forest: Karri, Marri and Jarrah." *John Austin Online*, 2017, accessed December 17, 2017, http://www.jbaphoto.com.au/australianforest.html.

41   John Austin. "Karri Forest Logging." *John Austin Online*, 2017, accessed December 17, 2017, http://www.jbaphoto.com.au/forestlogging.html.

42   John Austin. "Forest Protest and Activism." *John Austin Online*, 2017, accessed December 17, 2017, http://www.jbaphoto.com.au/forestprotest.html.

43   National Library of Australia. "Peta Sargison and Debbie Ludlam on the 'Megastump', Gardner 08 State Forest, Chesapeake Road, Northcliffe, Western Australia, 18 January 1999." *Trove*, 2017, accessed December 17, 2017, http://nla.gov.au/nla.obj-137020066/view.

and Activism" continues this theme with a catalogue of protest images, some of which take place in the forests and others at urban demonstration sites in Perth. For instance, "Dawn Vigil Lane State Forest, October 1988" is a moving interpretation of karri form that radically deviates from the sublime, picturesque, and grotesque traditions of karri art, which either exclude human activities more or less completely or centralize the gory impacts of saws, axes, tractors and clear-felling technologies. In this emotive composition, an assembly of tree defenders, clasping hands and forming a circle, surround a karri in which other activists sit in a platform high in the canopy.

In contrast to the activist ethos of Austin and the abstractions of Taylor, Tony Windberg's paintings and photography develop a more contemplative aesthetic that directly engages the organic materials of the sites he visits. Residing in Northcliffe, Windberg draws inspiration from the karri forests, as he explains in his documentary "A Painter in the Woods:" "I was walking through the bush and there was this big tree. I turned around, and there was this big wall of a karri tree. It was just one of those lovely moments, an epiphany if you like, where I thought 'I can see where I'm going with this'. The tree had all the colours and the bark that flaked off, and the sense of the cycles of nature. So I began going down that road of doing artworks based on trees."[44] The oil-on-canvas "Karri,"[45] painted between 1985 and 1990, approximates the color fidelity of a photograph. The image is exceptional for its intimate attention to the lowest portion—rather than the transcendent heights—of the karri. The vivid details of previous seasons' fires are evident in the charring of the trunk base and, as the bark decorticates, tawny colored flakes collect on the ground, producing a richly textured scene.

Windberg's paintings, "Harvest Red" and "Harvest Gold," incorporate a range of local plant materials, including karri charcoal and ash.[46,47] Each work consists of two rectangular panes, with the first work's left pane depicting a farming track through karri country with stacked logs in a paddock off in the distance. The right-hand pane offers a more obscure, heavily shadowed scene,

44   Gerne Mehr Films. "Tony Windberg: A Painter in the Woods." *Vimeo*, 2017, accessed December 17, 2017, https://vimeo.com/107500539.

45   Tony Windberg, Artist. "Harvest—Red" and "Harvest—Gold." *Tony Windberg, Artist, Online*, 2017, accessed December 17, 2017, http://www.tonywindberg.com/#!Karri/zoom/uzj99/dataItem-ikhk5h6f2.

46   Yahoo News 7. "Northcliffe Artist Lands Success (Pictures)." *The West Australian, Regional*, 2017, accessed December 17, 2017, https://au.news.yahoo.com/thewest/regional/south-west/a/16181260/northcliffe-artist-lands-second-survey-success/

47   http://www.tonywindberg.com/#!for-sale/rllgh.

perhaps at sundown. In distinction to the images of intensive logging that pervade the history of karri art, the atmosphere of "Harvest Red" is refreshingly one of what appears to be sustainable, small-scale, subsistence forestry. "Harvest Gold" is an analogue composition, also comprising two rectangular panes with the left-hand pane featuring the same country road and paddock but at a different season. A stalwart karri is a steadfast figure between both works, affirming the constancy of the great trees within the cycles of human and environmental change. Similarly, Windberg's "Woods/Wald" combines ash, charcoal, earth pigments, resins and acrylic binders on linen to create two parallel depictions of the same burnt karri.[48] Like "Harvest Red" and "Harvest Gold," the work focuses on one subject, a karri in this instance, but techniques, perspectives and effects—most obviously, coloration—vary between the two interpretations.

To be certain, a number of other contemporary artists have turned to karri trees and forests for inspiration. H.E. Quicke worked in the timber industry in the Manjimup-Pemberton area before taking up art in his retirement years. "Near Warren Camp"[49] is an astute rendering of the karri forests adjacent to the Warren River, a popular recreational spot along the Heartbreak Trail driving route. The image presents a tranquil scene of two people fishing from a tin boat as the karris loom overhead. As depicted by Quicke, who would have witnessed his share of forest clearing over the years, karri country is a place for relaxation, rejuvenation, conviviality, and contemplation. In a similar vein, Carol Thompson's "Karri Brook"[50] is an engrossing interpretation that accentuates the pale-pinkish hues of the karri boles as a brook passes below. Evocative of Windberg's "Harvest Red" and "Harvest Gold," Thompson's "Grey's Road"[51] focuses compositionally on the unsealed road through the forest with a paddock visible off to the right. The painting attends to the intersections between farmland and forest that are ever-present yet in perpetual flux as karri country continues to transform and evolve—affect and be effected—into the twenty-first century and beyond.

48   South Australian Museum. "Woods/Wald." *South Australian Museum Online*, 2017, accessed December 17, 2017, http://waterhouse.samuseum.sa.gov.au/gallery/2013/woods-wald/

49   Gold n' Grape Gallery. "Near Warren Camp—H.E.Quicke." *Gold n' Grape Gallery Online*, 2017, accessed December 17, 2017, http://goldngrape.com.au/?attachment_id=637.

50   Gold n' Grape Gallery. "Karri Brook." *Gold n' Grape Gallery Online*, 2017, accessed December 17, 2017, http://goldngrape.com.au/?attachment_id=737.

51   Gold n' Grape Gallery. "Greys Road." *Gold n' Grape Gallery Online*, 2017, accessed December 17, 2017, http://goldngrape.com.au/?attachment_id=726.

Conclusion

This broad, historical survey of karri art from European settlement to contemporary times elucidates the remarkable extent to which karris have inspired artistic production in Western Australia. The astonishing height and visual beauty of karris have always galvanized the efforts of painters, photographers, and print-makers since Marianne North through to those at work today. However, artists have also been captivated by the intimate spiritual and sensory qualities of the forests and not necessarily those qualities located in the lofty upper reaches of the trees. In recent years, the fragility of the great Southwest eucalypts, in the face of industrial logging, has come into focus within some creative interpretations. In broad terms, this chapter has highlighted the contribution of visual art to raising awareness and building appreciation of the environment, specifically inimitable old-growth natural heritage. Thus, a critical plant studies of old-growth forests would incorporate visual media in its analysis (see chapter 2). To be sure, conservation ideas, such as "photography for environmental sustainability" (Giblett and Tolonen 2012, 228), position visual art as a means to foster harmonious, long-term, mutually beneficial and, even, symbiotic relationships between people and nature, including the "vegetable giants of the west" at the heart of *Forest Family*.

Acknowledgments

The author wishes to acknowledge Fiona Sinclair of Southern Forest Arts for suggesting contemporary artists to include in this survey. Thanks also go to the Northcliffe Visitor Centre for their generous input.

Bibliography

Abbott, Ian. 1983. *Aboriginal Names for Plant Species in South-Western Australia*. Perth, WA: Forests Department of Western Australia.

Allbrook, Malcolm. 2014. *Henry Prinsep's Empire: Framing a Distant Colony*. Canberra: ANU Press.

Beard, John. 1990. *Plant Life of Western Australia*. Kenthurst, NSW: Kangaroo Press.

Boland, D.J., et al. 2006. *Forest Trees of Australia*. Collingwood, Vic: CSIRO Publishing.

Campbell, James Archibald. 1890. "A Naturalist in Western Australia." *The Australasian* April 5: 692.

Campbell, James Archibald. 1901. *Nests and Eggs of Australian Birds*. Sheffield: Pawson & Brailsford.

Campbell, James Archibald. 1921. "History of a Jarrah Board." *The Australasian* May 14: 868.

Chappell, Christabel. 2016. "Van Raalte, Henri Benedictus Salaman (1881–1929)." *Australian Dictionary of Biography*. Accessed March 25, 2017. http://adb.anu.edu. au/biography/van-raalte-henri-benedictus-salaman-8904.

Commonwealth of Australia. 2017. "Australia's Bioregions (IBRA)." Accessed March 25, 2017. http://www.environment.gov.au/land/nrs/science/ibra.

Crawford, Patricia, and Ian Crawford. 2003. *Contested Country: A History of the Northcliffe Area, Western Australia*. Crawley, WA: University of Western Australia Press.

Ednie-Brown, John. 1899. *The Forests of Western Australia and Their Development, with Plan and Illustrations*. Perth, WA: Perth Printing Works.

Ewers, John. 1946. "The Trout Comes to Western Australia. " *Walkabout* 13: 35–37.

Gascoigne, John. 2002. *The Enlightenment and the Origins of European Australia*. Cambridge, UK: Cambridge University Press.

Gaynor, Andrew. 2012. *Guy Grey-Smith: Life Force*. Crawley, WA: University of Western Australia Press.

Giblett, Rod, and Juha Tolonen. 2012. *Photography and Landscape*. Bristol, UK: Intellect Press.

Hallam, Sylvia J. 1975. *Fire and Hearth: A Study of Aboriginal Usage and European Usurpation in South-Western Australia*. Canberra: Australian Institute of Aboriginal Studies.

Heathcote, Christopher. 2002. "L'education Sentimentale." In *Australia Felix: Landscapes by Jeffrey Makin*, edited by Christopher Heathcote, 13–26. Melbourne: Macmillan Art Publishing.

Hopper, Stephen. 2004. "Southwestern Australia, Cinderella of the World's Temperate Floristic Regions." *Curtis's Botanical Magazine* 21.2: 132–80.

Hurley, Frank. 1953. *Western Australia: A Camera Study*. Sydney: Angus and Robertson.

Hutton, Drew, and Libby Connors. 1999. *History of the Australian Environmental Movement*. Cambridge, UK: Cambridge University Press.

Jones, Philip. 2011. *Images of the Interior: Seven Central Australian Photographers*. Kent Town, SA: Wakefield Press.

Lane, Cindy. 2015. *Myths and Memories: (Re)Viewing Colonial Western Australia through Travellers' Imaginings, 1850–1914*. Newcastle upon Tyne: Cambridge Scholars Publishing.

McEvey, Allan. 1979. "Archibald James Campbell (1853–1929)." *Australian Dictionary of Biography* 7. Accessed March 25, 2017. http://adb.anu.edu.au/biography/ campbell-archibald-james-5483.

North, Marianne. 1894. *Recollections of a Happy Life: Volume 2*. New York: Macmillan and Co.

Pike, A.F. 2017. "Hurley, James Francis (Frank) (1885–1962)." *Australian Dictionary of Biography*. Accessed March 25, 2017. http://adb.anu.edu.au/biography/hurley-james-francis-frank-6774.

Plate, Cassi. 2005. *Restless Spirits: The Life and Times of a Wandering Artist*. Sydney: Picador.

Sassoon, Joanna. 2012. "Poignant, Harald Emil Axel (1906–1986)." *Australian Dictionary of Biography*. Accessed March 25, 2017. http://adb.anu.edu.au/biography/poignant-harald-emil-axel-15467.

Seddon, George. 1997. *Landprints: Reflections on Place and Landscape*. Cambridge, UK: Cambridge University Press.

Snell, Ted. 1995. *Howard Taylor: Forest Figure*. Fremantle, WA: Fremantle Arts Centre Press.

Staples, A.C. 1988. "Prinsep, Henry Charles (Harry) (1844–1922)." *Australian Dictionary of Biography*. Accessed March 25, 2017. http://adb.anu.edu.au/biography/prinsep-henry-charles-harry-8119.

Von Mueller, Ferdinand. 1882. *Systematic Census of Australian Plants*. Melbourne: M'Carron, Bird & Co for Victorian Government.

Von Mueller, Ferdinand. 1887. *Key to the System of Victorian Plants*. Melbourne: Robert S. Brain, Government Printer.

Western Australia Bureau of Agriculture. 1897. *The West Australian Settler's Guide and Farmer's Handbook, Volume 1*. Adelaide: E.S. Wigg & Son.

CHAPTER 7

Sing the Karri, Sculpt the Jarrah: Sustaining Old-Growth Forest through the Arts

Robin Ryan

What names and stories
do they have for the mess of our doings?
This forest has given shelter
and done people in. Find one tree here—
look carefully and remember
that art can do no better.

FROM "TREES" BY LUCY DOUGAN IN *THE FOREST WAITS* (TAYLOR, DOUGAN, AND GILLAM 2006)

∴

Of Western Australia's Southwest landscape, Jan Taylor writes: "One is so dwarfed by the dimensions of this environment that one cannot help feeling that one's true place is merely an insignificant part of a huge biological whole" (Taylor 1990, 130). Rod Giblett's history of settler relations with the natural systems of old-growth forests (chapters 4 and 5) illustrates the ways in which the forests are valued for economic, environmental, cultural, and social imperatives. John Ryan's discussion of the artistic inspiration of the region's signature karri (*Eucalyptus diversicolor*) tree in chapter 6 advances the consideration of how plants *act upon us*, contributing to the co-generation of cultural practices (Ryan 2012a, 104). The current chapter targets the interface between music, sculpture, and these iconic trees in and around a small town within the Shire of Manjimup. Geographically situated forty-four kilometers south of Manjimup, Northcliffe is known for its deeply layered history of natural glory and ruin.[1]

---

1 Northcliffe sits within the Warren Biogeographical Region and the Mediterranean Forests, Woodland, and Scrub Terrestrial Ecosystem (Australian Government Department of the Environment and Energy, "Australia's Bioregions—Maps") (See also the beginning of chapter 6 in this volume).

© KONINKLIJKE BRILL NV, LEIDEN, 2018 | DOI 10.1163/9789004368651_009

FIGURE 7.1   *"Karri Trees at Northcliffe."*
IMAGE COURTESY OF ROBIN RYAN.

Broadly defined, forest arts ecology relates to, reflects, and relies on nature. Through integrating a study of art and music with natural and social arboreal ecology, I will articulate a sense of how artistic practices, processes, and products of nature reawaken respect for old-growth forests—of how, through embodied environmentalism, we might address global responsibility for conservation. In raising ecological awareness, musical compositions and visual artworks inform public understandings of sustainability, for as "nature's *otherness* inheres in us [italics in original]" (Watkins 2007, 21), we are charged with the responsibility of sustaining it. We cannot ignore the real-politik of the environmentalist mandate because many forest populations around the globe are threatened, including in the Southwest of Western Australia.[2]

---

2 The greenhouse effect is directly related to plants (effective hoarders of carbon). Burning fossil fuels and forests releases carbon dioxide into the atmosphere, leading to increased retention of heat in the atmosphere and oceans, a consequent climate change, and rise in sea levels. The solution must involve forests since fewer trees means less moisture entering the atmosphere, i.e. more droughts. An awareness of global warming is propelling the creation of art, music, and literature aimed to close gaps in environmental understanding. This expanding context for working in, with, or around nature sees contemporary artists balancing their rhapsodies about (perceived) pristine places with ecocritical representations of the unruly nature of constantly changing environments.

A number of sculptures will be understood to foster close connections with trees, based on the premise that to work in symbiosis with them is to highlight the powerful agency of *Eucalyptus*, a genus which, according to Hay (2002, 221), emerged approximately eighty-million years ago in Australia's fossil record. Undisturbed in their natural habitats, eucalypts were, in Hay's words, "as essential to the survival of other flora and fauna, to the preservation of soil and the provision of clean water as they were to the preservation of the scenic value of wilderness or as a unique genetic repository" (Hay 2002, 201).

The Southern Forest Sculpture Walk, dedicated by the non-profit organization Southern Forest Arts (henceforth SFA) in 2006, was the first purpose-built walk trail within Australia to permanently feature specially commissioned artworks from a range of different art forms. In reflecting the social, cultural, and environmental heritage of the Southern Forests region, the 1.2 kilometer walk circuit adjacent to the Northcliffe Visitor Centre weaves its way through some majestic karri, jarrah, and marri forest. In this foundational space of inspiration and mystification, artistic and literary media express the relationship of humans to the natural world of old-growth ecosystems.

SFA rebranded the Walk "Understory: Art in Nature" in 2009, followed by "Understory Art in Nature Trail" (henceforth "Understory," or "the Trail"), in 2015. The *understory*—a layer of vegetation beneath the forest's *overstory* or canopy—is symbolic of Northcliffe's chequered caché of personal and communal memories. A strong sense of the spirit of forest is projected in the body of literature commissioned by SFA and, for the purposes of this chapter, the album *Canopy: Songs for the Southern Forests*.

SFA Project Coordinator Fiona Sinclair has authorized and generously assisted this study. It draws on personal experiences of the Trail in summer, autumn, and spring; an aural study of *Canopy* informed by Robyn Johnston's liner notes; an undated *Trailguide*; and the perspectives of various musicians and sculptors. The unpaginated *Southern Forest Arts Sculpture Walk Catalogue* (henceforth *Catalogue*) provides useful information on the music and sculptures as well as the poetry, prose, and storytelling components of Understory.

Following a brief account of tall forest history, I will describe the establishment of the Trail. By my third visit to Northcliffe in March 2016, I had begun to intuit connections between the music and the visual art. My focus on the inspirational role of karri, jarrah, and marri as conduits for cultural creation thus compares twelve sculptures with twelve musical items highlighting the agency of forest flora (see Appendices 1 and 2). Together, Understory art and *Canopy* music sustain country with acts of seeing, listening to, and interpreting the forest; of remembering its past and present importance to Indigenous Australian people; and of recalling the battles of early settlers with the insurmountable terrain. The creative media discussed below highlight some of the poignant aspects of this history documented by J.P. Gabbedy (1988), Patricia and Ian Crawford (2003), and Carole Perry (2014). Northcliffe lies in the heart of the karri belt, and karri remains the distinctive theme of its cultural histories.

Tall Forest History

Indigenous people and settlers have known forest in different ways but as Clarke and Johnston (2003, 3) note, Western constructs do not adequately account for the depth of Indigenous cultural relationships to land. In their comprehensive study of Northcliffe, Crawford and Crawford (2003, 1) note that the Murrum sub-group of the Nyoongar nation began to inhabit the lower Southwest at least fifty-thousand years ago. According to Nyoongar belief, Ancestral Beings created the Southwest's principal trees. The fact that karri and jarrah are strongly associated with Women's Dreamings implies that women have played a major role in the traditional management of forests (Crawford and Crawford 2003, 25). Intimate Indigenous association with Northcliffe and its hinterland extends back at least six-thousand years (SFA 2006, *Catalogue*), a meta-narrative reflected in Understory's Indigenous art, literature, and music.[3]

Interleaved with these examples are cameos of settlers who moved beneath the canopy in dialogical engagement with the trees (see chapters 4 and 5 of this volume). Northcliffe's foundation in 1924 was based on Premier James Mitchell's belief that wherever the karri tree grew, good pasture could thrive once the trees were removed. As the broadcaster and historian Bill Bunbury (2015, 24) puts it, "Few at the time realized that the rich-looking brown loam, while it nourished karri, was not capable of creating lush English meadows." Following the abandonment of some farms during the 1930s Depression, the mid-1940s brought soldier settlers to the district to boost dairy production and found a tobacco industry. But it was the advent of woodchipping that changed the nature and intensity of the destruction of karri ecosystems (Crawford and Crawford 2003, 192).

Fast-forwarding to the 1990s and the turn of the millennium, the Southwest community experienced major social, political, and economic upheaval during the old-growth forest debate between conservationists and the timber industry (SFA 2009, "Understory Art in Nature").[4] The successful protest conducted at

---

3   It was not until three years after the Trail opened that Ecotourism Australia began to promote ecologically sustainable tourism with a focus on natural areas that foster environmental and cultural understanding, appreciation, and conservation. A core agenda is respect and acknowledgment of a region's Indigenous history (Eastwood 2009, 77–8).

4   Group Settlement farming schemes led to massive deforestation for little economic or social return. The largely unrestricted logging practices that followed were perceived to be harmful

Giblett Block (described by Nandi Chinna in chapter 8) was followed by protests, blockades, camps, and shifts in and around the Northcliffe vicinity. Media
attention was drawn to the town as protesters chained themselves to tree
trunks and suspended themselves from branches. Continued conflict led to
the signing of the Western Australian Regional Forest Agreement in 1999.

The Genesis and Construction of the Sculpture Trail

The deregulation of the dairy industry in 2001 saw a sharp decline in the area's
population. With many remaining families still harboring resentment over the
forest debate, Sinclair conceived a vision of a special forest in which the arts
would provide a platform for positive change. The notion of artists working
organically and sustainably to reveal the beauty of natural dimensions would
have appealed to Western Australia's first Forest Conservator, Charles Lane
Poole (1885–1970), as he was sensitive to the aesthetics of forests:

> Lane Poole felt that when people had actually reached what he called
> 'forest consciousness', then they would also arrive at a position where
> they believed that forests need to be used responsibility [sic] and
> sustainably.
>
> MICHAEL CALVER, INTERVIEWED BY BILL BUNBURY, JULY 28, 2014, CITED IN
> BUNBURY 2015, 145[5]

With experience as an ephemeral artist working for the Wilderness Society inside threatened forest coupes, Sinclair initially envisaged a sculpture trail deep
within the pristine heart of Northcliffe Forest Park's old-growth karri stand.
Conceding to the reaction this idea would provoke from "the most fluorescent
of the Australian Greens," she modified her ideas, and in 2005, presented a dis-
creet alternative to the Manjimup Council, highlighting potential employment
opportunities for Northcliffe residents through the project (Fiona Sinclair, interviewed by the author, September 26, 2014).

Sinclair won her pitch, and the Council leased a twelve-acre pocket of allo-
cated Crown Land as an A-Class Reserve, vested for Education and Recreation.
While the Trail only occupies a tiny pocket of the beautiful Northcliffe Forest
Park, its flora represents a microcosm of the broader forest and its lively world

---

to the region's ecology, and appeals were made to the Federal Government for a legislative
resolution.

5   Hay (2002, 202) claimed that only 10% of Australia's pre-settlement old growth forest
remained.

of non-human agency: the richly diverse understory of smaller trees and shrubs including karri oak, boronia, hazel, and wattle; creeping vines and tiny orchids; and the animals and birds inhabiting the forest.[6]

The SFA committee drilled down to the project's core via lengthy consultations to secure unified community ownership. Local music teacher Dave Carrie proposed that a music component be included, and local writers suggested stories and poetry. SFA raised over six-hundred-thousand Australian dollars, employed a public consultant, and selected mentors for each creative mode. In addition to a survey of flora and fauna, the hygiene of all soil types was examined. Limestone was brought in from Witchcliffe and shale from closer to town. However, protocols stipulated that no biomass could be removed from the forest and that predominantly raw, natural materials were to be used. Trail supervisor Roy Moss directed forest structure and earth moving, and for environmentally sensitive reasons, backhoes were used instead of cherry pickers (Fiona Sinclair and Peter Hill, interviewed by the author, September 26, 2014).

With *forest* as prescribed image (wider than the bounded project site), SFA commissioned a team of regional, urban, and internationally acknowledged artists, musicians, and writers to consider five themes (Table 7.1) as a basis for work that was to be "fresh, intimate, and grounded in place." Cautious about foregrounding political dimensions in the project, Sinclair promoted bioregional identity in her description of the trail as "a love letter from those of us who live here to our visitors, to connect with grace" (*pers. comm.*, April 6, 2014). For the sake of reuniting a divided community, the visual art, music, poetry, and prose components were to respect—with unembellished rhetoric—the differing associations between people and forest, but as will become evident, a degree of historical tension surfaces in a few of the sculptures.

Aware of the corrupting impact of tourism, the commissioned sculptors committed themselves to the creed of the Artists in Nature International Network (AiNIN) that art in nature implies respecting nature, not using or abusing it for the sake of art:

> We think that this respect for nature implies creating a specific work for each site, as a way of revealing and commenting upon our relationship with, our environments. We feel that "art in everyday environments" is

---

6   Karri country provides a haven for bees and lorikeets harvesting pollen and nectar, and for emu (*waitch*), kangaroo (*yonga*), wallabies, quokkas, and tamars (traditional owner Suzanne Kelly in SFA *Catalogue*).

TABLE 7.1   *Southern forest sculpture walk: themes prescribed for artists*

Spirituality	Creativity	History	Dichotomy	Sensory
sacredness, sanctuary, forest as cathedral, meditation, mystery	fertility, energy, life cycles, renewal, beauty, inspiration	geological, ecological and human (from early Indigenous to contemporary)	contrasts of nature (light and dark, growth and decay), diversity, paradox	sights, sounds, smells, tastes, and touch of the forest

SOURCE: HTTP://WWW.UNDERSTORY.COM.AU/ART.PHP.

a positive way to connect with a large part of the population who, so far, have not been interested in the traditional 'object oriented' presentation of art in the removed environments of museums, galleries and public spaces. Art in nature is an area of art where a 'new public' can at last be engaged. Many artists and communities in the world have been creating and progressing along these lines over the past few decades. Their efforts (our efforts) have brought an increasing consciousness of this approach to art making to the population at large.

AiNIN 2016, "MISSION"

Each participant spends time in the area exploring and developing ideas for sculpture vis-à-vis a range of aesthetic ideals, compositional materials, and sculptural styles and techniques that reflect the flexibility of contemporary art in interaction with the organic elements of nature, and human social history. SFA commissioned Kevin Draper, who works with forged steel (classed as organic), to create an entry statement and trail head sculpture. Regarding the Walk as a kind of "window" into aspects of the local environment, Draper visually "floated" the architectural form within the branch and leaf shapes of the symbolic trees. The colors of the frame respond to the natural environment by referencing fire, plant forms, and bird life (Draper, in SFA 2006, *Catalogue*). Given the unbounded and often unpredictable nature of fire and weather, several pieces are classed as "ephemeral works in process of decay." The result is a cultural landscape built on its reception by viewers as much as the inspirations of artists (Figure 7.2).

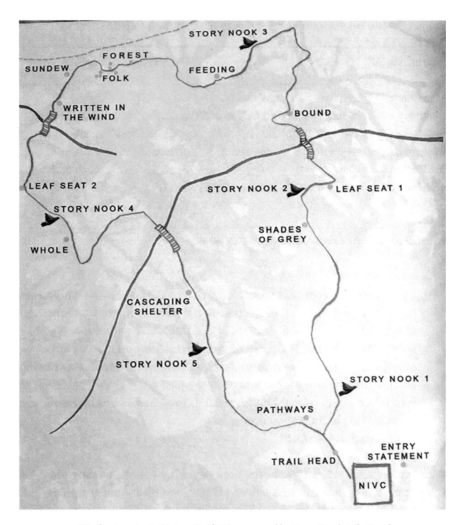

FIGURE 7.2 *"Understory Art in Nature Trail." Map created by Fiona Sinclair for* Southern
Forest Sculpture Walk Catalogue *(2006).*
IMAGE COURTESY OF SOUTHERN FOREST ARTS.

Capturing the Melody of *Canopy*

It is plain to see how visual art arranges itself around the Trail, but how
does humanly organized sound inscribe the site? All of the musicians par-
ticipated in forest workshops at the observation point, and of the five MP3
audio tours edited and mixed by Graham Evans to enable visitors to dialogue
with the bush, Audio Tour 3, *Songs for the Southern Forests*, references tracks

featured on the album *Canopy* with personal reflections on their sources of inspiration.[7]

Speaking of the process of "looking for nature right in the music" and "music right in the midst of nature," musician and eco-philosopher David Rothenberg (2001, 7) has observed that "the melody and the rhythm can bring the message alive, keep it moving and current, and the best of this music transforms the message into something that can be carried no other way." To help move the town past the divisiveness of the turn-of-the-millennium forest campaigns, the artists shaped forest presence laterally. Sinclair describes this form of implicit environmentalism as an "around the back" approach that avoids lapsing into political commentary or judgment (*pers. comm.*, April 6, 2014).

Under producer Lee Buddle's direction the musicians supported each other's musical discoveries. Appendix 1 indicates how they selected instruments for the descriptive ability of their tone color (timbre) to draw out environmental and cultural referents of place. Like the artworks, the musical offerings described below may be regarded as intrinsically bioregional to the lower Southwest's ecology, endemic fauna and flora, and sociocultural identities. Thematically the twelve songs focus on nature, people, and human-nature relationships including cultural histories, personal memories, and emotional insights promoting appreciation of forest. Northcliffe's geographical distinctiveness, thus, illustrates a notion fostered by the co-founder of Common Ground, Sue Clifford, that "a place can be read as a story; that nature and culture have come up with something unique in a place and full of meaning for the people who live there" (Clifford in Busch 2007, "Poetic Justice").

In the opening track "My Place" (2006), local Windy Harbour musician Ann Rice draws on childhood memories of her family's farm near Pemberton. Rice recalls her father's descriptions of giant karris falling to the sound of the cross-cut saws and bushmen's axes that foresters used before the advent of chainsaws and log trucks. It is a theme echoed in the sculpture "Feeding" (2006, metal rod, gold leaf), for which visiting French artist François Davin created an original story about the big trees:

> It had to do with the awe, love, fear, and even the hatred that these trees have provoked among the settlers. It had to do with their heroic determination, especially of the mothers to raise families in a difficult environment. It also took into consideration the 'fall' of many trees.
>
> SFA *CATALOGUE*

---

7   All of the musicians performed at The Walk's official opening on November 25, 2006, with an estimated 1,000 people in attendance (SFA *Catalogue*).

Artistic advocacy promoting sustainable forests and the extent to which plant forms exert an aesthetic role in Understory thus deserve close notice.

Trees as Art and the Art of Trees

In a critique of the Trail published by *Artlink* in 2007, Thelma John suggested that "the problem with art placed in the forest is that the forest is so perfect already that the art can easily feel presumptuous, superfluous, definitely imposed, in some cases even invasive" (John 2007, "Northcliffe Sculpture Walk"). Indeed, since the natural world may exceed our ability to represent it (Watkins 2007, 21), art does well to leave out more than it includes. The goal of art is thus to highlight, or amplify, the "specialness" of what is already there.[8] As subjects manifesting their own forms of gesturing, the trees may need little intervention but an artist's focus brings to our notice things that might otherwise go unnoticed such as the immense variety of organic form, color, contour, patterning, light, and shade along the Trail.

SFA attempted to include artworks that belong *in* the forest and are *of* the forest, works such as the strikingly architectural "Cascading Shelter" (Tony Pankiw, 2006), a retainer in the form of birds' wings, leaves, and other forest shapes that directs rainwater back to the resident fauna. People have a right to their own response, and Thelma John found a tension in the shelter's placement in the forest. On a contrasting note, she found the subtle interventions into the forest's existing phenomena to be "quietly sublime" (John 2007, "Northcliffe Sculpture Walk"). Of her work "Bower" (2008), Gemma Dodd explained, "I'm inspired by forms made by little creatures who use instinct to create—manipulating humble materials" (SFA *Trailguide*).

Lightness of touch is exemplified in the small branches that Cecile Williams stapled to a tree trunk with fine twigs to create a pattern representing "Women Meeting" (2006). Williams also collected banksia leaves and twigs to create a line tracing the light falling on the decaying matter of a log and the underbelly blackness of where the tree was burnt ("Banksia Line," 2006, ephemeral work in process of decay). Four years later, the artist salvaged and repaired an abandoned cut marri log with sticks ("Log Life," 2010).

Space for discovery and reflection varies as the seasons furnish different temporal experiences of the cycles of nature. During a warm summer hour

---

8 With the aim of liberating people to become "overtly passionate about their connection with nature, life and place," Sue Clifford and Angela King founded the organization Common Ground in the UK in 1983. They encouraged artists to express the specialness of their surroundings (Clifford in Busch 2007, Website).

exploring the Trail on January 5, 2014, the freshness of the unmediated bush magically absorbed my senses. The trees forms began to emerge as sculptures in their own right, besides which art literally exuded out of their bark, twigs, and leaves—perhaps because I entered the site with the catchword "art in nature" in mind? As my locus of attention began to drift between a plant world that had evolved over millennia and the immediate offerings of the contemporary art world, I was surprised to notice a mysterious bark plane hovering in the understory. Since this was not a commissioned work, Sinclair suggests that it was either created by an *in situ* workshop student or a visitor, or was a just a case of "nature really having fun with your brain" (*pers. comm.*, July 23, 2016).

Indigenous Regeneration of Space

Just as particular trees emanate special auras, the artworks fashioned around them function as receivers through which we can pick up energetic patterns or impressions of ancient forest. The ability of nature to reveal the diminutiveness of the human being comes into focus in the dozen *Forest Folk* (2006) that sculptor Richie Kahaupt created from marble resin, cement, iron dust and steel to lurk amongst undergrowth built layer on layer over time. These meditations of an unusual sort raise pointed questions. They symbolize people who, in the words of Kahaupt, "have been reduced, pushed down, held down by the weight of history" (SFA 2006, *Catalogue*).

As Rod Giblett and Hugh Webb in their chapter "Living Water or Useless Swamps?" point out, the Indigenous landscape is not a mere "object" for representation; in fact "lands are the sites—in some cases the *meanings*— of very important cultural stories" (Giblett and Webb 1996, 5). Thus country is not a touristic scene, but a life-sustaining habitat (John Ryan 2012b, 290). The Murrum-Nyoongar people experienced the lower Southwest forest in a cycle of seasons that followed the annual rhythms of nature. For ritual purposes, men adorned themselves with colored feathers collected from the great variety of Southwest birds. The Pibelmun (Bibbulman) people of the Manjimup area accompanied their extempore songs and "dream songs" with the boomerang (*kylee*), throwing stick (*meero*), and handclapping or beating of the club (*dowuk*) on the ground (Robin Ryan 2003, 32). The visual and acoustic communities of forest trees undoubtedly contributed to this cultural pulse.

To fulfill her commission from SFA, artist Norma MacDonald (of Yamatji and Nyoongar descent) created "Written in the Wind" (2006, Figure 7.3) in two-dimensional form to represent the strength and power of the forest as it

FIGURE 7.3   *"Written in the Wind" by Norma MacDonald (2006).*
              IMAGE COURTESY OF LUISA RYAN.

symbolizes Indigenous links to the region. As she searched for a space among the giant trees for her sculpture, MacDonald (in SFA 2006, *Catalogue*) was overwhelmed by the realization that her people once walked proudly through this land:

The welding of my sculpture is symbolic of the piecing together of the past history of the Pibelmun people. The black steel rod structure on the karri tree is contrasted against its white trunk, a reminder of the contrasting cultures of earlier times. Standing tall and proud in the centre of the work is an Aboriginal man. He stands on the rock and becomes 'one' with the tree and the forest: as it was for thousands of years before European settlement and as it will always be.

The artist remained aware—in her attempt to weave time and place together—that time will change the vegetation around her sculpture, and that her creation will change with its surroundings as time goes by.

In 2004, the Nyoongar Yorga woman Della Rae Morrison of the Bibbulmun and Wilman nations teamed up with Murri musician Jessie Lloyd (Gugugimidhir tribe of Cooktown, North Queensland) to form the multi-award winning duo "Djiva." The duo's 2006 contribution to Understory encapsulates the fundamental qualities of balanced forest ambience in the song "Ngank Boodjak" (literally "Mother Earth"):

This was inspired by walking the trail down at Northcliffe, feeling the energy of the land and the beautiful trees and hearing the birds. When I find a spot that I love, I try to feel out the lay-lines, which feel like vortexes of energy coming out of the ground; it's pretty amazing.

MORRISON, in CD SLEEVE

The vibrational power of seed force is explained in Stanza 1 of "Ngank Boodjak":

The wind carries the seed
Where the seed falls, there remains forever
a vibration in the ground
It's called Djiva

The final stanza laments the disappearance of traditional life in the forest melodically:

No land or belonging
Noongka boodjak
Mother Earth she cries
Ngank Boodjak waliny

The addition of the punctuation "Ssh!" draws out the fundamental cultural importance of walking slowly and listening intently to "beautiful Mother Earth."

Morrison describes her Elders' belief that when we leave this life from our physical bodies "our spirit is earthbound and is living in the rocks or the trees and if you listen carefully you might hear their voices and maybe you will get some answers to your questions" (SFA 2006, *Catalogue*).[9] The speaking and singing of Indigenous language is believed to harness energy (vibration) in country. Its inclusion in "Ngank Boodjak" challenges mainstream constructions of forest, thus widening the scope of Crawford and Crawford's historical discussion around the incommensurability of Indigenous and settler cultures, and, more specifically, of how European interference with existing landscapes diminished vital Indigenous knowledge of how to live in balance with forest.[10] Since the "natural charisma" with which karri inspires the medium of painting (chapter 6) also extends to the music, literature, and sculpture made about karris, I discern certain reverberations between these media in the interart analysis below.

"King Karri" Well Sung

We mostly sing the praise of karri's visual worth. With an upright, knot-free trunk that can tower to ninety meters (two-hundred-and-ninety-five feet), the species has famously drawn sightseers to reside permanently in Northcliffe. Using an otherworldly vocal timbre, Bridgetown-born musician Joel Barker suggests that the mighty old-growth tree has an *inside* as well as an *outside* story by allowing it to speak in the first person. The first and last verses of his song "Something for Everyone" (2006) honor karri's longstanding survival, thereby connoting its role as a more-than-human witness to past and present forms of Indigenous livelihood in the forest:

> Perhaps I am old
> See the wisdom in my bones
> I have seen the brightest days
> And darkness beyond my age
> I stand the test of time
> I'm the laughter in your eyes
> I'm the future because of my past
> I'm the memory meant for every life

---

9   In an e-mail to the author on July 15, 2014, Morrison explained how, because country is so precious to her, she co-founded the Western Australian Nuclear Free Alliance to campaign against the uranium industry.

10   "Ngank Boodjak" may also be heard on Djiva's debut album, *Yowarliny* (2009).

Assumedly this depiction also encompasses the turbulent chronicle of white settlement. The notion of karri bringing "laughter" resonates with the sub-theme of Anna Jacobs' commissioned novella *The Group Settler's Wife* (2006) in which, as an antidote for a harsh life, the heroine seeks consolation and refreshment in her weekly walks through the beautiful forest. Likewise Barker's lyric "I'm shelter for your kin / I'm the place where love begins" resonates with the romantic plot of this novella.

Rod Giblett's references to "king karri" and the timber industry's "yields of pleasure and profit" (chapter 4) are echoed in the song's vital chorus:

> There's something for you
> Like there's something for him
> And I'll try to give it to you
> I'm the King

Despite Barker's image of karri's utilitarian ability to provide for all walks of life ("people take what they need from the tree, be it for money or for conservation"), karri is not as industrially useful as jarrah (chapter 4). Nevertheless its structural hardwood provides long lengths of timber: "I'm the money for your loan / I'm protection for your home." For me, Barker's musical personification of old-growth karri as a king with a (canopy) crown becomes clear to the eye in the sculpture "Whole, You Were Meant to be Here" (Lorenna Grant, 2006). I describe the creative construction of this commissioned artwork below to exemplify the productive stages of inspiration/stimulus, design, preparation, incubation, and installment of forest art.

A Sculpture up in the Canopy

Lorenna Grant's work represents an expanding inquiry into the creation of sculptures based on experiential moments where art, nature, and humanity fuse, and her website "Lorenna Grant—Artist" describes ephemera and site-specificity as her primary motivations. A regard for natural/biological systems is reflected in this and other examples of Grant's award-winning art that give form to the unseen forces that move through land, water, and air. Noticing that the other artists were confining their work to the understory, Grant was motivated to produce a sculpture that would be visible at the level of the canopy.

On becoming aware of the aura (energetic field) of a tall *karri tree*, the artist envisioned a torus (a large convex molding shaped like a ring doughnut) encircled around its upper bole.[11] The karri itself could be conceptualized as the supporting "actor" or "actant" having agency within the artwork. Grant collected naturally occurring fibers (small branches and strips of bark shed by the host tree and other karris) and spent eight days working directly beneath the tree. She wove the organic materials into two separate halves of a stainless steel frame, and wired these substructures together around the base of the karri to form a nested halo (*pers. comm.*, July 5, 2016; see Figure 7.4).

Hoisting the completed torus 13 meters up the bole of the karri into a position where the torus could "hover" or "float" in seamless suspension was SFA's greatest challenge. Local builder Ross Facius and Trail supervisor Roy Moss employed a bow-and-arrow shooting technique to advance and secure the first ropes across the upper branches. With what Grant describes as "strenuous work executed with organized ease," Facius and Moss proceeded to put more ropes up the tree using classic tree-climbing devices, including abseiling.

Grant described the moment of instalment as "organic and amazing as people who had been "big players" in the rescue of the Boorara karri forest started appearing out of the bush in collective support of the project" (*pers. comm.*, July 5, 2016). Three teams amounting to twenty volunteers hauled ropes and pulleys in different directions to hoist the artwork up the tree so that Facius and Moss could connect it to stainless steel cabling with steel suspensions and fixings (fine cabling fixed to the karri trunk does not damage the tree).

For some, the regular shape of the elevated torus (Figure 7.5) may impose itself too strongly on the irregularly shaped landscape. However its measured presence only augments the artist's malleable message of transformation and protection. Grant has come to regard the artwork as a "generative, natural maternal shelter or home for creatures such as birds or insects to live in." The fact that "the torus vibrates along with the tree" (*pers. comm.*, July 5, 2016) leads me to conceptualize this artwork as another way of "singing the karri." Although intentionally ephemeral, the torus has maintained its structural integrity and presence over the last twelve years.

One of Grant's intentions was to make the invisible visible—to represent the toroidal fields that surround life, from the tiniest atoms to the largest planets in our solar system. The weighty physical structure projects energy and power as it opens itself up for a stream of associations, depending on a viewer's background or cosmological belief system. Grant, for instance, sees an "auric field" (pers. comm., July 5, 2016). In the SFA 2006 *Catalogue* she describes the work as an act of communion that is central to all relationships:

---

11   In geometry, a torus is a surface of revolution generated by revolving a circle in three-dimensional space about an axis coplanar with the circle.

FIGURE 7.4 *"Aerial View of Lorenna Grant with Her Completed Torus, October 2006."*
IMAGE COURTESY OF SOUTHERN FOREST ARTS AND ROSS FACIUS (HTTP://LORENNAGRANT.COM).

FIGURE 7.5 *"Close-up View of 'Whole, You Were Meant to be Here' by Lorenna Grant (2006)."*
IMAGE COURTESY OF LUISA RYAN.

Within its design and surface texture, I've woven my responses to the forest and the many relationships we have with it. Its shape is that of a torus—a circle—a field of energy—the wind—a nest—a platform—a halo—a ring. Viewers may consider notions of the ring or circle as a symbol of completion, of love—and in loving.

Relationships between people, and between people and forest, likewise permeate the torch song featured on the *Canopy* album.[12] Jazz/blues singer Libby Hammer recounts a passing romance in which "the waiting forest listens and believes" in organic communion with the song's subjects. Beginning with a harmonic loop on guitar, "The Glade" (2006) portrays the forest's intermittent light and rain, beauty, solitude, and airy space. A resonant double bass speaks of old-growth forest; a cello expresses suppleness and renewal; and drums imitate the insect community's polyrhythmic undercurrent. A bird-like soprano saxophone punctuates the singer's plaintive questioning of the hill, the reeds, the trees, and the wind as she rehearses unrequited love (paraphrased from SFA 2006, *Canopy* sleeve notes).

Mirroring the Marriscape

As John Ryan points out in chapter 6, "karri ecology will endure in the spaces between human and non-human communities." Marri commonly grows on sand inside the karri forest where it hosts a rich understory of edible food plants. Northcliffe resident Warwick Backhouse enjoys savoring the grain, color, and texture of native timbers. He carved the seats "Calophylla" and "Marginata" (2006) from locally sourced jarrah and marri, and strove to balance the character of their respective leaf shapes with comfort for sitting.[13] *Calophylla* (the marri genus) means "beautiful leaves," a reflection on marri's luxuriant foliage of broad, shiny dark green leaves with a paler underside. Using space symbolically, Backhouse positioned the seat within a grove of marri.[14] The fibrous, flaky trunks of these trees can grow to fifty to sixty meters

---

12   A torch song is a sentimental song in which one party has become oblivious to the existence of the other.

13   The jarrah *Marginata* seat was replaced in 2017 on account of storm damage.

14   Marri is known as "red gum" because its grey to dark brown trunk exudes the dark red *kino* that Nyoongar people ingested for medicinal purposes. Marri's woody urn-shaped fruits (honkey-nuts) famously inspired artist May Gibbs to create the legendary storybook characters in "Tales of Snugglepot and Cuddlepie" (1918).

(one-hundred-and-sixty-four to one-hundred-and-ninety-seven feet), and they are easily blackened by fire.

Art and Fire

It was inevitable that the fires that shape Northcliffe Forest Park would be reflected in Understory art and song. To represent the reality of regional bushfire, Natalie Williamson wrapped her pink metal "Sundew" sculpture (2006) around a burnt-out stump, and painted the leaves red to mimic flames. This whimsical work plays with scale by enlarging flowers of the tiny carnivorous native sundew (*Drosera menziesii*) so that its subtle beauty becomes apparent. The sculptor hopes that spiders will spin their webs about the artwork to blend the boundaries of her representation of a flycatcher with real flycatchers (spiders) (SFA 2006, *Catalogue*). In another example of art incorporated into the landscape, Graham Hay's return of more than twenty-thousand pages of (mostly unread) government reports to the forest ("Nurture II," 2008), came with the intention that "impregnated with native scarlet bracken fungus spores, these digits may grow and have a life of their own" (SFA *Trailguide*).

The shape and texture of Duke Albada's "Shades of Grey" (2006, superglass corrugated sheet, steel, and resin) is inspired by the "statuesque karri as it sheds its bark" (SFA 2006, *Catalogue*) in annual revealings of its creamy-white to fawn underflesh. The thought-provoking sculpture is an ode to the local people that asks the visitor to find a balanced viewpoint in the "greys" between "black and white" opinions about fire in the forest, while red patches of resin represent the "overwhelming importance of fire management" (SFA *Catalogue*). On a visit to Understory on September 27, 2014, I noticed that the charred vegetation from a recent controlled ("cool") burn looked strikingly beautiful following a heavy spring shower. It spoke a language of opportunity for nature to reassert itself as seedpods burst and continue the cycle.

In his song "White Haze" (2006), cabaret performer Tomás Ford recounts the scene of a week-long cool burn. Having experienced the tranquil scenic drive into the town via Pemberton, Ford envisaged the smoke as a kind of metaphor for the beautiful mystical healing nature of Northcliffe:

> Smell steaming karri leaves
> Lost by lonely trees
> Forest never grieves
> Feel no flames licking here
> No pain could get near
> Just glowing white air

Boosting the skills of state and interstate firefighters, volunteer Peter Hill (born in Manjimup, resides in Northcliffe) and a team of local men spared both Understory and the town from a colossal conflagration in early 2015.[15] In the difficult aftermath of this traumatic event, SFA mounted an *After the Burn* photographic exhibition, followed by *Out of the Ashes*, comprising paintings and woodwork produced with ash, charcoal, and resin. Combining both exhibitions, *After the Burn—On Tour* reached wider audiences at Bunbury regional galleries between December 2015 and February 2016. *Rising From the Ashes*—a series of sculptures created to commemorate the 2015 bushfire—includes cast faces of Northcliffe and Windy Harbour residents by artist Kim Perrier. The exhibition was launched at the Act Belong Commit Southern Forest Arts Festival held from November 26–27, 2016 to celebrate the tenth anniversary of Understory and the establishment of the Northcliffe Information and Visitor Centre.

Bare upper branches of karri usually indicate the presence of past fires, while new shreds of bark hang down like potential firebrands. Jarrah, on the other hand, has deep roots that allow the species to draw water from great depths. Its large underground lignotuber (swelling) stores carbohydrates and allows young trees to regenerate after fire. On that note, I turn to the sculptural importance of jarrah in Understory.

Jarrah as a Sculptural Resource

Known as Swan River mahogany in colonial days, jarrah became Western Australia's principal hardwood harvested for timber,[16] and its qualities of durability and workability also render it eminently suitable for creative sculpturing (chapter 4). A lush, deep red-brown before weathering to grey, the strong, fibrous bark of jarrah sheds in strips and is deeply striated with longitudinal fissures. Local builders Ross Facius and Paul Owens used sections of this drought-, termite-, and water-resistant species to construct the boardwalks that lead through the Trail's differing vegetation types (heathland, acacia country, tall forest stands, and riverine environments with three creek crossings).

The starkness that the parameters of size and structure can evoke is evident in the imposing work "Bound" (2006, cast bronze, jarrah, copper cable, rope, and steel) by Alex and Nic Mickle. To make a socially realistic statement about the battle for the forests (1992–2001) that had split families down the middle,

---

15   These local vigilant residents continue to push their equipment to the limit around the perimeter of Understory.

16   Jarrah was WA's first major export and a huge contributor to its economy.

the pair of artists viewed hours of video interviews with those "bound" to the forest on both sides of the blockades and protests. They then worked with a tapered section of karri trunk that had split through the center. Expanding on this symbol of divisiveness over the destruction and defense of forest, they crafted a bronze heart embedded in timber form, to display signs of struggle (cables, rope, and banding; see Figure 7.6).[17]

Thematically, plants jostle and struggle with each other just as people do. Drawing on practical knowledge of this phenomenon, Peter Hill welded a series of tall ladders to erect between some Understory trees in 2008. On March 19, 2016 the sculptor personally explained how this work "Competitive Ground" (composed of found sticks, steel, and paint) represents the plants of the forest in constant struggle for survival as they reach out and compete against each other for sunlight, nutrients, and water.

In his earlier work "Pathways" (2006), Hill had fused the characteristics of assorted natural forms in the Northcliffe region. Crafted from the timbers of jarrah and blackbutt (*Eucalyptus patens*, known as "yarri") and welded steel, branches reach out, linking the earth to the sky and the sky to the earth, mimicking the forest that surrounds it: "These pathways could be interpreted as a walk trail drawn over the land's surface or the branches of a tree that reach out into space to collect light" (Hill, in SFA 2006 *Catalogue*). The viewer is left to decide whether the sculpture is a tree or a flower, a fungi, or a microscopic form.

Other artworks added to the forest since 2006 have exploited jarrah. For instance, "Australia House" (Cornelia Konrads, 2008) poses the questions: "A house that is stretching out towards the tree? Or is the tree pushing down the house? Are they supporting each other, or is it a struggle? Happy or sad, they seem to be linked in a relationship—perhaps like us and Nature?" (SFA *Trailguide*). "The Golden Tree" (2012) is a tall found jarrah stump that Thomas Heidt (2012) tinged with 24-carat gold leaf to represent the American poet Walt Whitman's saying that the very flesh of those who give charity shall be "a great poem" (SFA *Trailguide*).

Soundscaping the Forest

The sensory modality of hearing is a valid source of ethnographic information for comprehending the innate musicality of forest. The pioneering soundscape artist R. Murray Schafer (1977) discovered how sounds produced by natural

---

17   On September 28, 2014, the author witnessed participants from both sides of the forest debate join in the parade to celebrate Northcliffe's 90th Birthday.

FIGURE 7.6   *"Bound' by Alex and Nic Mickle (2006)".*
IMAGE COURTESY OF LUISA RYAN.

acoustic behaviors enabled mediatory language between listener and environment. Recently, sound ecologist May-Le Ng (2014, "Listening to the Land") updated the definition of *soundscape* to "a collection of all the sounds in the environment, including man-made, biological, and geophysical sounds."

This concept of soundscape materializes in the *Canopy* album's organic/machine crossover of fused eco-jazz/trip-hop. Pete and Dave (Lo-Key Fu) Jeavons meld synthetic and natural sounds of the forest—how they work in harmony, and the way they clash. The pair recorded forest sounds on a mini-disc. They sampled the wind rustling through trees at nearby Mount Chudalup, the granite rock explained in Nyoongar accounts of the ancestors (Crawford and Crawford 2003, 19).[18] This passive opening wind in their item "Sanctuary" (2006) gradually builds to focus on the human-made elements, thereby avoiding a struggle between the organic and the mechanical. The brothers described this process to Robyn Johnston as one of "morphing and changing" (SFA 2006, *Canopy* sleeve notes).

When we use found objects to improvise with sound, we locate ourselves geographically to produce a freer, conceptually different type of music. The brothers banged logs together, and banged rocks against logs, to compose their main snare groove. By holding the natural and the human in tension, "Sanctuary" encapsulates the forest as a living, breathing entity and a refuge to which people can escape the pressures of the modern world. In the words of the Jeavons brothers, it provides "a way of reconnecting with our original selves" (SFA 2006, *Catalogue*).

Dunn (2001, 107) notes how codes of communication, like insect sounds, arise from the unique organization of living things. Rothenberg (2014, 2) suggests, moreover, that it is the rhythms of insects that bind us to the landscape, describing them as:

> One small sense that ties us to the eternal, for like all animal sounds, they have been around for millions of years longer than anything human. And the most important thing about them is that they may be the very source of our interest in rhythm, the beat, the regular thrum.

Using this trope as a basis to consider musical naturalism in Understory, I turn to an instrumental composition by David Pye (b. 1958).[19] It demonstrates the ecomusical agenda: one of listening to the subtle, sensory messages of the auditory environment, of allowing the site itself to determine the characteristics of musical composition.

---

18 Mt Chudalup overlooks plains that once hosted Nyoongar ceremonies (Suzanne Kelly in SFA *Catalogue*).

19 Pye founded Nova Ensemble in 1983, and in 2002 he formed the ensemble *pi* to focus on the use of strings, reeds, and percussion in an improvisational context.

Structuring "Cicadan Rhythms"

For his 2006 *Canopy* commission, Pye focused on the mass pulsing of cicadas alight in the karri forest flora. He tracked individual cicadas through the Trail as they sang, flew, and landed on treetops. Eventually finding a patch of low-lying scrub where he could listen to four or five individuals, the composer notated their distinct rhythms. This particular species—however many—were sounding together *in tempo* with individual rhythm patterns that interlocked to create one fantastic rhythm (Australian Broadcasting Corporation 2009, interview with Andrew Ford).

In this moment, Pye realized that the cicada chorus is the "summer soundmark" of the southern forests.[20] Male cicadas (*Homoptera: Cicadidae*) communicate *in plenum* (full assembly) by means of airborne sound (intraspecific signaling) in continuous trains of sound pulses that mask the forest's other natural sounds.[21] Through this natural phenomenon—the loudest known love song in the insect world—we become privy to a more-than-human way of communicating and a whole new understanding of the karri forest which Barker sings up as "the place where love begins."[22]

To take the listener through a "geographical linear representation" of the Trail, Pye sampled sounds of instruments made of wood to enhance the rich forest imagery (violins, cellos, clarinets, woodblocks, and temple blocks) and positioned these against layers of pre-recorded ambient sounds (birds, cicadas, frogs, and rain). The composer's stroke of brilliance was his inclusion of high-pitched Indonesian *angklung* instruments, for which he set patterns to represent the metronomic clicking of the cicadas.[23] Pye describes how he mapped

---

20   A "soundmark" is a community sound that is specially regarded or noticed by people in that community (Truax 1978, *Handbook for Acoustic Ecology*). Understory projects good acoustic definition because its natural "acoustic community" (Truax 2001, *Acoustic Communication*) creates a unifying relationship within the ecosystem.

21   A female cicada responds to a male with a wing flick, to which the male replies with more clicking. As the duet continues, he begins a courtship call. At other times he may shriek when startled but the mechanism for sound production remains the same. Large species produce calls in excess of 120 decibels at close range (*pers. comm.* from zoologist David Young, August 31, 2014).

22   Each species performs a distinctive song for attracting females. Since species presence might overlap in the forest during summer, one would need to collect specimens to establish precise identification.

23   *Angklung* consists of joint bamboo tubes suspended within a bamboo frame. The tubes are whittled and cut to produce certain notes when the frame is shaken or tapped. Each angklung produces a single note or chord, so several players collaborate to play melodies. The black bamboo used to build angklung is harvested during the two weeks a year when

Understory's microenvironments (e.g. low scrub, large trees, and sludgy creek) structure the walk and form the basis of the music:

> I walked around it with a stopwatch and noted how long it took to get through each section of the forest, and that became the musical timing of the various parts of the work. I was concerned to try and capture the textures of the forest, particularly the sounds, and also the play of light as the sun strikes down through the trees.
>
> *Canopy* CD SLEEVE

Tangled textures represent the thick tea tree-based understory at the start of the trail; more open, peaceful textures depict the Walk's move into the karri and sheoak sections of the far slopes; and watery sounds evoke the creek crossings. The climax Pye achieved with glockenspiels and bells depicts a serendipitous moment during a rainy winter walk: "I came upon a grasstree that was covered in very small drips of water. The sun suddenly struck them and they looked like diamonds scattered all over the tree" (David Pye, e-mail message to author, September 10, 2014). In its portrayal of Understory's energetic world of non-human agency, "Cicadan Rhythms" models the type of evocative nature composing that can engage listeners with a forest's age-old musical idiom.[24] It demonstrates how the mysterious agency of more-than-human sound can reflect the fullness of environmental health (and, by deduction, the dangers of environmental degradation).

Like insect sound, stone tells of fathomless geological age and primal connection with landscape. Formed from granite, bronze, and timber, Kati Thamo's "Forest Stone Series" (2006) incorporates sandblasted imagery. Thamo conceived the ecological theme in collaboration with writer Dianne Wolfer. The audio tour recording of five stories by Wolfer encourages six to eleven year-olds to explore the mysterious minutiae of the forest: "the secret unseen world of the forest dwellers—living in the undergrowth, the canopy and amongst the leaf litter" (SFA 2006, *Catalogue*). Interconnected by the presence of a central villain in the form of a feral cat, the stories provide a non-didactic way for children to consider sustainability and the impact of introduced species on the forest (derived from SFA 2006, *Catalogue*).

---

cicadas sing, and it is sustainably cut at least three segments above the ground to ensure the root continues to propagate (UNESCO 2009, "Intangible Cultural Heritage").

24   In 2009, Pye rescored "Cicadan Rhythms" for large orchestral forces. On September 6 he conducted its world première along with Grainger's orchestral masterpiece *The Warriors* (1913).

Cameos of Early Northcliffe Settlers

The circularity represented in Lorenna Grant's torus surfaces again performatively in the "endless loop" of Cathie Travers' "Lament" (2006). Compositionally, her silvery-toned accordion paints a picture of silver-green space inside large cathedrals of karri. Support singers weave wordless vocals around Travers' sparse spoken/sung text and contribute string improvisations. The images and sensations—real and imaginary—depict how Northcliffe was settled: "Two rooms, earth floor / corrugated iron door." A one-line tale is told of impoverished locals salvaging what they could from a shipwreck one Christmas, before the lament reaches its sad conclusion: "The child in her arms / she walks into the sea."

Nocturnal experiences of massive trees, moon, and stars led Brendon Humphries and Kevin Smith to evoke timelessness in their song "When the First Wind Blew" (2006). The pair fantasize the "untouchability" of a maiden moment (where no human exists) in the forest: the wind's initial movement through brooks and leaves. This idyll of Aeolian vibration (wind-induced oscillation) undergoes abrupt change as Smith recalls the hardships of his pioneering ancestors in the isolated little community: "the man's on fire / as we carry him /to the waterside."

The luminous role that music plays in commemorating Northcliffe Settler history culminates in the items by prominent Perth musicians Bernard Carney and David Hyams. "The Destiny Waltz" (Carney, 2006) is a tribute to some three-thousand group settlers who migrated to Western Australia with a promise and a dream of owning their own dairy farms, only to find that they had to clear huge karri trees with their bare hands and build their own furniture from kerosene tins and gelignite cases. Carney captures the sense of light relief offered by weekend dances held in one-teacher schools to popular tunes like "The Destiny Waltz" (played on the ill-fated maiden voyage of the *Titanic* in 1912). The item fades to a few haunting strains of this waltz by the Victor Military Band (1914) to "pay tribute to the era where the inspiration of the song came from" (Carney in SFA 2006, *Canopy* sleeve notes).

It is the energy of Celtic rhythm that illuminates a Settler imagination in David Hyams' instrumental medley. The introduction to "Awakening" evokes the morning sun streaming through majestic tall trees before the music segues into a waltz played on an accordion. The item recalls the role this instrument played in dance music before the Settlers were in a position to buy pianos.

Hyams' concept for "Shaking the Tree" (2006) is a lilting Irish jig representing humankind's struggle with forest in strong winds and rain. The atmospheric theme resonates with the CD's earlier track "Shelter" (2006) in its depiction of Northcliffe's wet climate. Mel Robinson's rich string harmonies describe the shifting environment of a brewing storm that fortuitously led her to seek help and become acquainted with local residents. In a more optimistic sounding jig, Hyams' third and final title, "When the Light Comes," defers to the saying by American conservationist John Muir that "the wrongs done to trees, wrongs of every sort, are done in the darkness of ignorance and unbelief, for when the light comes the heart of the people is always right" (quoted in SFA 2006, *Canopy*, sleeve notes).

Conclusion

This chapter investigated the creative practices, processes, and products inspired by the confluence of nature and culture in Northcliffe's Understory Art in Nature Trail. The staging of art in a forest raises questions, and puts itself in question, but it is evident that Understory is a place where sculptural and musical agency emerges as a sustaining (as opposed to a destroying) force for the ecological and cultural values of local flora. Each medium stands on its own merit while a degree of integration emerges through the artists' varied utilization of karri, marri, and jarrah (Appendix 2). Jarrah, in particular, furnishes a substantial medium for artistic expression, exemplified in works such as "Australia House," "Pathways," and "The Golden Tree."

Like the forest flora, the sculptures exhibit degrees of change and continuity. Some composed materials even blend together with natural elements over time. The more ephemeral works leave an impression of the natural fragility of forest and the benefits of gentle human impact. Beliefs and practices influence ways in which people value and interact with forest, while art and music express the stories and concerns of culture. The more "permanent" works play discursive roles, evidenced in the profound on-site spirituality of Norma MacDonald's monument to her ancestors' terrain on a karri tree. The torus that Lorenna Grant created in the understory was hoisted up into the canopy to represent protection and transformation. In both these works, karri as the supporting actor is far more than a mere backdrop to art.

A variety of natural and human references are held in tension, as different cultural readings of the karriscape are valued. The album *Canopy* incubates aesthetic appreciation of forest in its creation of varied musical atmospheres. Several musicians express characteristic human experiences of karri, and in Joel Barker's song, karri assumes its own discursive identity. Two tracks employed the phenomenological approach of experiential sensitivity to the auditory environment and the specters of arboreal, biological, or cultural loss. Djiva worked with the land as vibrating material in a Nyoongar language "singing up" of forest that repositions and revitalizes the Indigenous human within karri country. Composer David Pye ingeniously transcribed unmuted "Cicadan Rhythms": the lovelorn voice of a single breed of creature in the karri forest ecosystem.

These acts of "singing the karri" and "sculpting the jarrah" suggest ways in which our voices might blend in more graciously—and more respectfully—with a karri forest that opens our hearts to its grandeur. The establishment of this unique creative hub propelled Northcliffe's move toward a healthier phase of community evolution in which Southern Forest Art's invitation for an interdisciplinary response is ongoing in its efforts to fulfill the ecological thematic of sustainability. Local luminary Fiona Sinclair led a large team of creatives whose exploits at the cutting edge of ecology tapped into broad inventories of art, literature, and sound. Theirs is an exportable model that proves the ability of regional arts to link social heritage to the knowledge of natural biodiversity—in best practice, with the humility not to over-intrude.

Acknowledgments

It is a pleasure to thank Fiona Sinclair, Peter Hill, Ann Rice, Lorenna Grant, Robyn Johnston, Della Rae Morrison, Carole Perry, David Pye, Aline Scott-Maxwell, and David Young for research support; Kumi Kato for piquing my interest in forest arts ecology; and Jennifer Yeoman and Noelle Johnston for Northcliffe hospitality.

Appendix 1

Music Composed for Canopy: Songs for the Southern Forests (*prod. Lee Buddle, 2006*)

Composer/genre	Track title / theme	Instrumentation
1. **Ann Rice** (folk storytelling)	**"My Place"** (memories of family, farm and forest)	Vocals/guitars/accordion
2. **David Pye** (percussive musical naturalism)	**"Cicadan Rhythms"** (interpretation of insect and forest sounds)	Angklung/violin/cello/ woodblocks/ temple blocks/clarinet/tapes

Composer/genre	Track title / theme	Instrumentation
3. **Mel Robinson** (up-tempo indie)	**"Shelter"** (running away from a storm)	Vocal/cello/double bass
4. **Djiva** (contemporary Indigenous song)	**"Ngank Boodjak"** ("Mother Earth")	Vocals/acoustic, electric & slide guitars/drums/percussion
5. **Cathie Travers** (improvisatory collage)	**"Lament"** (for the Settlers' struggles)	Accordion/vocals/guitar/piano/violin/drums/programming
6. **Brendon Humphries Kevin Smith** (country rock with spoken word)	**"When the Wind First Blew"** (nature was there before people)	Vocals/guitars/dobro/drums/piano/percussion
7. **Libby Hammer** (jazz/blues torch song)	**"The Glade"** (remembering love in the forest)	Vocal/guitar/soprano sax/cello/double bass/drums
8. **Pete & Dave Jeavons** (fused jazz/trip-hop/electronica)	**"Sanctuary"** (forest as refuge; melding natural and human sounds)	Guitars/percussion/talking drum/cowbell/soprano sax
9. **Tomás Ford** (cabaret punk rock)	**"White Haze"** (fire management; healing)	Vocal/programming/guitar
10. **David Hyams** (medley of Celtic-infused soundscapes)	**"Awakening"** / **"Shaking the Tree"** / **"When the Light Comes"** (forest, Settlers and storms)	Guitar/mandolin/dobro/bodhran/rainstick/cello/accordion/flute
11. **Bernard Carney** (folk storytelling)	**"The Destiny Waltz"** (Settlers' dance for relief)	Vocal/guitar/accordion/drums/recording of "Destiny Waltz"
12. **Joel Barker** (up-tempo indie rock)	**"Something for Everyone"** (karri speaks in the first person)	Vocal/guitars/percussion

SOURCE: CD SLEEVE NOTES BY ROBYN JOHNSTON; HTTP://WWW.UNDERSTORY.COM
.AU/ART.PHP.

Appendix 2

Sculptures Inspired by, Composed of, or Attached to, Jarrah, Karri or Marri

Artist's name / date	Title / theme	Compositional materials
1. Alex & Nic Mickle (2006)	"Bound" (interpreting the battle for the forests)	Cast bronze, jarrah, copper cable, rope steel; split karri trunk
2. Cornelia Konrads (2008)	"Australia House" (support and struggle)	Jarrah, nails and paint; sculpture leans against a lopsided tree
3. Thomas Heidt (2012)	"The Golden Tree" (inspired by Whitman's thoughts on almsgiving)	Found jarrah, tinged with 24-carat gold leaf
4. Peter Hill (2006)	"Pathways" (branches linking Earth to sky)	Jarrah and blackbutt timber, welded steel
5. Norma MacDonald (2006)	"Written on the Wind" (symbolizes Indigenous links to the region)	Steel, steel rods and epoxy paint; a two-dimensional structure attached to karri
6. Lorenna Grant (2006)	"Whole, You Were Meant to be Here" (a communion; energy field)	Steel, karri sticks and wire; a torus suspended high around a karri tree with fine steel cabling
7. Warwick Backhouse (2006)	"Calophylla" (mirroring the leaf shape of marri)	*Corymbia calophylla* (marri seat situated amidst marri trees)
8. Cecile Williams (2010)	"Log Life" (repairing an abandoned cut log)	Marri log salvaged from location and found sticks
9. Peter Hill (2008)	"Competitive Ground" (plants compete for sunlight, nutrients and water)	Found sticks, steel and paint; ladders erected around several tall trees
10. Cecile Williams (2006)	"Women Meeting" (pattern represents women meeting; ephemeral work)	Found sticks stapled to a fallen tree trunk with fine twigs

SOURCE: SOUTHERN FOREST ARTS UNDERSTORY *TRAILGUIDE*; HTTP://WWW.UNDERSTORY .COM.AU/ART.PHP

Bibliography

Artists in Nature International Network (AiNIN). 2016. "Mission." Accessed November 14, 2016. http://www.artinnature.org/mission.

Australian Broadcasting Corporation (ABC). 2009. "Composer David Pye." Interview with Andrew Ford. *The Music Show*, Radio National, September 12. Accessed January 30, 2015. http://canadapodcasts.ca/podcasts/MusicShowThe/1225021.

Australian Government Department of the Environment and Energy. "Australia's Bioregions—Maps." Accessed 14 November, 2016. http://www.environment.gov.au/land/nrs/science/ibra/australias-bioregions-maps.

Bunbury, Bill. 2015. *Invisible Country. South-West Australia: Understanding a Landscape*. Crawley, Western Australia: UWA Publishing (Charles and Joy Staples South West Region Publications Fund).

Clarke, Annie, and Chris Johnston. 2003. "Time, Memory, Place and Land: Social Meaning and Heritage Conservation in Australia." *Intangible Heritage*, ICOMOS 14th General Assembly and Scientific Symposium, Victoria Falls, Zimbabwe, October 27–31. Accessed January 26, 2015. http://www.international.icomos.org/victoria-falls2003/papers/B3-7%20-%20Johnston.pdf.

Busch, Simon. 2007. "Poetic Justice." *Communities* interview with Sue Clifford and Angela King, *The Guardian*, Wednesday December 19. Accessed June 17, 2016. https://www.theguardian.com/society/2007/dec/19/communities.voluntarysector.

Crawford, Patricia, and Ian Crawford. 2003. *Contested Country: A History of the Northcliffe Area, Western Australia*. Perth: University of Western Australia Press.

Dunn, David. 2001. "Nature, Sound Art, and the Sacred." In *The Book of Music and Nature: An Anthology of Sounds, Words, Thoughts*, edited by David Rothenberg and Marta Ulvaeus, 95–107. Middletown, CT: Wesleyan University Press.

Eastwood, Ken. 2009. "At One With Nature." *Australian Geographic: The Journal of the Australian Geographic Society* 96 (October-December): 72–83.

Gabbedy, J.P. 1988. *Group Settlement. Part 1. Its Origins, Politics and Administration. Part 2. Its People: Their Life and Times: An Inside View*. Nedlands, WA: University of Western Australia Press.

Giblett, Rod and Hugh Webb. 1996. "Living Water or Useless Swamps?" In *Western Australian Wetlands: The Kimberley and South-West*, edited by Rod Giblett and Hugh Webb, 1–9. Perth, WA: Black Swan Press/Wetlands Conservation Society.

Grant, Lorenna. "Lorenna Grant—Artist." Accessed July 1, 2016. http://lorennagrant.com.

Hay, Ashley. 2002. *Gum: The Story of Eucalypts and Their Champions*. Sydney: Duffy and Snellgrove.

Jacobs, Anna. 2006. *The Group Settler's Wife*. Northcliffe, WA: Southern Forest Arts.

John, Thelma. 2007. Review of "Northcliffe Sculpture Walk." *Artlink: The Word as Art* 27.1. Accessed April 15, 2014. https://www.artlink.com.au/articles.cfm?id=2936.

Ng, May-Le. 2014. "Listening to the Land." *Ecos: Science for Sustainability*. CSIRO Publishing, December 12. Accessed December 15, 2014. http://www.ecosmagazine.com/?paper=EC14294.

Perry, Carole. 2014. *Northcliffe: The Town That Refused to Die*. Albany, WA: Digger Press.

Rothenberg, David. 2001. "Introduction: Does Nature Understand Music?" In *The Book of Music and Nature: An Anthology of Sounds, Words, Thoughts*, edited by David Rothenberg and Marta Ulvaeus, 1–10. Middletown, CT: Wesleyan University Press.

Rothenberg, David. 2014. *Bug Music: How Insects Gave Us Rhythm and Noise*. New York: Picador.

Ryan, John C. 2012a. "Passive Flora? Reconsidering Nature's Agency through Human Plant Studies (HPS)." *Societies* 2: 101–121. doi: 10.3390/soc2030101.

Ryan, John C. 2012b. *Green Sense: The Aesthetics of Plants, Place and Language*. Oxford: Trueheart Academic.

Ryan, Robin. 2003. "Aboriginal Traditions: People of the Southwest." In *Currency Companion to Music and Dance in Australia*, edited by John Whiteaok and Aline Scott-Maxwell, 32. Sydney: Currency Press.

Schafer, R. Murray. 1977. *The Tuning of the World*. New York: Knopf.

Southern Forest Arts (SFA). 2006. *Canopy: Songs for the Southern Forests*. CD project co-ordinator Fiona Sinclair; funded by the Australian Government under its Regional Partnerships and grants for Forest Communities programs. Recorded and produced by Lee Buddle, sleeve notes by Robyn Johnston. West Perth: Sound Mine Studios. 94558.1.

Southern Forest Arts. 2006. *Southern Forest Sculpture Walk Catalogue* (unpaginated booklet). Northcliffe, WA.

Southern Forest Arts. 2009. "Understory—Art in Nature." Accessed 12 April, 2016. http://www.understory.com.au/.

Southern Forest Arts. Undated pamphlet. *Trailguide. Understory Art in Nature Presented by Southern Forest Arts*. Northcliffe, Western Australia.

Southern Forest Arts (ed. Fiona Sinclair). 2016. *After the Burn: Stories, Poems and Photos Shared by the Local Community in Response to the 2015 Northcliffe and Windy Harbour Bushfire*. Second edition. Northcliffe, WA: Southern Forest Arts.

Taylor, Andrew, Lucy Dougan, and Kevin Gillam. 2006. *The Forest Waits: An Anthology of Poems*. Northcliffe, WA: Southern Forest Arts.

Taylor, Jan. 1990. *Australia's Southwest and Our Future*. Kenthurst, NSW: Kangaroo Press.

Truax, Barry, ed. 1978. *Handbook for Acoustic Ecology*. Vancouver, BC: ARC Publications.

Truax, Barry. 2001. *Acoustic Communication*. 2nd edition. Westport, CT: Ablex Publishing.

UNESCO. 2009. "Intangible Cultural Heritage. Indonesian Angklung." Accessed 28 June, 2016. http://www.unesco.org/culture/ich/en/RL/indonesian-angklung-00393.

Watkins, Holly. 2007. "The Pastoral After Environmentalism: Nature and Culture in Stephen Albert's Symphony: *RiverRun*." *Current Musicology* 84 (Fall): 7–24.

CHAPTER 8

Old-Growth Activism: The Giblett Forest Rescue of 1994 and 1997

Nandi Chinna

FIGURE 8.1 *"Chris Lee at Giblett Forest Rescue." 1997.*
IMAGE COURTESY OF BETH SCHULTZ.

© KONINKLIJKE BRILL NV, LEIDEN, 2018 | DOI 10.1163/9789004368651_010

The karri tree and the karri forest, it's one of the most beautiful trees in the world. And the forest itself, particularly the primeval forest, what we call the virgin forest, it had a cathedral like structure, somewhere where you feel entirely relaxed and the kind of the place you would always want to come back to. That's the way it affects me.

JACK THOMSON, FORMER FALLER/FORESTER (QTD. IN BUNBURY 1983)

You get in the middle, deep in the forest and you look around you and you just can't imagine how the rest of the world can behave as it does.

RON MELDRUM, FORMER FORESTER (QTD. IN BUNBURY 1983)

You feel that it's a living thing that has gone. You felt something was dying every time you saw a tree fall.

KATHLEEN FFOULKES, TIMBER TOWN RESIDENT (QTD. IN BUNBURY 1983)

In 1997 I was living and working in the tiny former mill town of Quinninup, situated in the Southwest of Western Australia between Pemberton and Manjimup. I moved to this part of the country, where, nearly one-hundred-and-fifty years earlier, the Giblett family had made their homes, built farms, and driven their cattle through the karri forests to the coast because of the extraordinary beauty of the forests of the surrounding area (see chapters 2 and 3 of this volume). It was not long before I became aware of the thin façade of this wild place. Behind the majestic roadside strips of jarrah (*Eucalyptus marginata*), marri (*Corymbia calophylla*, also known as *Eucalyptus calophylla* R. Br.), and karri (*Eucalyptus diversicolor*) lie the clear-fell forest coupes scoured like gaping wounds in the landscape. Large-scale clear-felling of native forests was in full swing and a vocal movement of forest conservationists was organizing to meet the challenge of protecting the old-growth Southwest forests from further destruction. The debate over the future of Western Australia's unique forest heritage, set against the clear-felling and woodchipping practices of the timber industry, came to a head first in 1994 and again in 1997 at a forest block named Giblett.

During the 1990s, Giblett forest block (hitherto referred to as "Giblett") became the site of one of Western Australia's first long-standing forest protest blockades, an action which was to change public perceptions of forests and forestry practices, instigate previously unheard of public participation and consultation in forestry management issues, and, to some degree, contribute to the electoral loss of the incumbent Court Liberal government. The Giblett forest rescue is seen by some participants and observers (Robertson 2015; Schultz 2015) as a symbolic event which has resulted in profound impacts upon the people involved, as well on as subsequent successful community campaigns in Western Australia. The principles and techniques of non-violent resistance developed during Giblett continue to be utilized in campaigns such as the James Price Point Gas Hub Action of 2007–13, in which the local community, Traditional Custodians, independent scientists, environmental groups, and some politicians successfully mounted a vigorous, creative, and effective on-the-ground resistance to protect the Kimberley from a proposed gas hub (Robertson 2013); and the "Love Makes a Way" campaign which is a faith-based movement seeking an end to Australia's inhumane asylum seeker policies.

Pivotal to the success of the Giblett Forest Rescue was the philosophical and practical ethos of non-violence, which was the basis of every action and decision (Lee 2015; Maddock 2015). This chapter uses interviews with participants and archival resources to discuss the community campaign to protect Southwest forests, and the environmental, social, and political consequences of the blockade of Giblett.[1]

Situating Giblett

Giblett forest block is situated four-hundred kilometers (two-hundred-and-forty-nine miles) south of Perth and about fifteen kilometers (nine miles) north-west of the town of Pemberton in the Southwest of Western Australia, and is now incorporated into the Greater Beedelup National Park. Giblett contains a mixture of old-growth karri and karri/marri plant communities with a small area of jarrah and jarrah/marri. There are also some very old blackbutt (*Eucalyptus todtiana*) trees and most of the understory of Giblett supports ancient sheoak groves.[2]

Giblett Block is named after the Giblett family, an overview of which is provided in chapters 4 and 5 of this volume. John Giblett (1809–82) and his wife Ann (1818–96) arrived in Western Australia on the ship *Simon Taylor* in 1842. In 1863, John Giblett took over a property at Manjimup and later moved

---

1   For a detailed history of the conservation movement in WA, see Ron Chapman (2008).

2   Old-growth forest was defined by the 1992 National Forest Policy Statement (NFPS) as "forest that is ecologically mature and has been subjected to negligible unnatural disturbance such as logging, road building and clearing" (Department of Conservation and Land Management, 2000). This definition focused on forest in which the upper stratum or over-storey was in the late to over mature growth phases. Then, in 1997, the Joint Implementation sub-committee (JANIS 1997) developed this definition further, describing old-growth forest as that which is "ecologically mature forest where the effects of disturbance are now negligible" (ibid).

to Balbarrup. The Gibletts played a significant role in the life of the fledgling community of the Manjimup area. They built a small school and paid for a school teacher to teach the local children, and held church services at their home, shouldering the expenses for a minister to travel to Manjimup a few times a year to hold services. Thomas Giblett later built a church at Dingup (Sclater 2000).

The Giblett family connection to the forest environment lies in their working lives in the ecotone between the jarrah and marri forests. In 1904, Walter and Hubert Giblett felled a huge karri tree on the edge of the Donnelly River and used its massive trunk to construct a bridge known as One Tree Bridge (see chapters 4 and 5). The Gibletts cleared jarrah forest to establish their farming properties, and Channybearup Road, which connected the Giblett properties at Manjimup with their pastoral leases on the coast, was originally named Giblet (*sic*) Road (Sclater 2001). This Giblett cattle route, now named Channybearup Road, runs close to the Giblett forest block.

The original nomadic custodians of the district where the Giblett family settled are the Nyoongar (Murrum) people (see chapter 3). For over fifty-thousand years, the Murrum have lived in the area, shaping the landscape through their firing and cultivating practices (Crawford and Crawford 2003), using the land as a source of food, and developing management techniques to ensure the continuation of their food resources. The Murrum were working the land on a sustainable yield basis long before the term was coined and became *de riguer* (South West Development Commission 2012). Due to their light imprint and nomadic way of life, the Murrum have left little built evidence of their presence in the area but their culture is held in *boodjar* (land), *moort* (family), and *katitjin* (knowledge and story), and continues to be shared among current generations of Nyoongar people (Collard 2007). The Walgenup Aboriginal Corporation has been active over a number of years in collecting oral histories of local Indigenous community members to preserve some of the stories of the area. There are several registered Nyoongar heritage sites close to the town of Manjimup, including wetlands, artefact scatters, and a stone arrangement at Dingup. While the significance of this stone arrangement is not known, its presence reiterates that the country where the Gibletts settled was already known and inhabited by Nyoongar people.

In *Fire and Hearth: A Study of Aboriginal Usage and European Usurpation in South Western Australia* (1975), Sylvia Hallam writes about the use of fire by traditional Nyoongar people to modify the landscape, one aspect of which was to allow for movement through the karri and jarrah forests. Her research shows that some areas of forest were burned while other areas were left dense and were used less (Hallam 1975). Fire was used to germinate tree seeds and stimulate the growth of grasses. The Aboriginal system of bestowing plant and animal totems also ensured the protection of species for future generations (Stocker, et al. 2015). Traditional ecological knowledge (TEK) systems include seasonal and species-specific understanding. The way Nyoongar people moved through the country, inhabiting different places during different seasons, also served to allow the regeneration and replenishment of flora and fauna species (Nannup 2003; Stocker, et al. 2015). The Manjimup Shire gains its name from the indigenous *manjin* reed, whose edible roots were highly valued by Nyoongars.

Giblett Block first gained public attention in 1974, when the Conservation through Reserves Committee was established by the Western Australian Environmental Protection Authority (EPA) to review WA's conservation reserves. The adjacent Beedelup National Park had been gazetted State Forest in 1901 and was changed to National Park in 1915. In 1974 additional area was sought to add to the relatively small Beedelup National Park. Three options were considered: Beavis, Strickland, and Giblett blocks. Strickland was chosen, but the values of all of the old-growth forest in the area, including Giblett, led it to be interim-listed on the National Estate Register of the Australian Heritage Commission (Sclater 2001). In 1994, the Western Australian department of Conservation and Land Management (CALM) and the Australian Heritage Commission (AHC) placed all of Giblett forest on the interim list of the Register of the National Estate for its wide range of conservation values. This listing, however, did not protect Giblett in 1994 and again, in 1997, when attempts to log the old-growth forest of Giblett Block led forest activists to occupy the site.

Forest as an Economy

The felling of trees as markers of colonial intent began in Australia with the first arrivals of British people into Cadigal country (Sydney Cove) in 1788. A work party was sent ashore to set up camp on the banks of the fresh water stream that was to become known as Tank Stream. Of the occasion, the colonial administrator David Collins observed "the stillness of which had then, for the first time since the creation, been interrupted by the rude sound of the labourers axe and the downfall of its ancient inhabitants" (Collins 1789). In Western Australia, the founding of Perth was marked by the felling of a tree at a spot near present-day Perth Town Hall (Moore 1884).

The harvesting of native jarrah and karri hardwoods was one of the first major industries in the Southwest. During his 1827 exploratory voyage to

Western Australia, Captain James Stirling commented upon the potential commercial usage of the plentiful timber supplies. On a trip to Geographe Bay he wrote, "wood is abundant here for the use of ships" (qtd. in Cameron 1981), and suggested the Asian market as a possible destination for WA timber. Later, in 1838, he observed that karri forests contained wood "of the finest description for ship building" (qtd. in Statham-Drew 2005). Chapman (2008) asserts that early colonial attitudes towards native forests reflected both a nostalgia for English landscapes and the desire to exploit forests as a resource. Heathcote supports this assertion, arguing that by 1870 the dominant attitude toward native forests was as a commodity for commercial exploitation (Heathcote 1976). Between 1788 and 1914, Powell estimates that the effect of colonization on the landscapes of Australia had been profound, with widespread ecosystem loss and a deeply entrenched attitude of exploitation of natural resources (Powell 1976).

Prior to 1890, the extreme isolation and lack of transport infrastructure in the Southwest region was prohibitive to the development of agriculture and other industries. Apart from tin mining at Greenbushes and coal mining at Collie, exploration for other mineral wealth had proved fruitless. Bryce Moore writes that post-1890 planning decisions regarding Southwest Australia centered upon soil (agriculture) and forests (timber) (as discussed in Moore 1993).

The M.C. Davies Karri and Jarrah Timber Company was a timber empire that hosted hundreds of employees and laid over one-hundred kilometers (sixty-two miles) of private railway, and even built its own private ports for exporting timber (Hamling 1979). A network of railway lines linked the Southwest region's many small timber mills to the Busselton Jetty, where native timber was loaded onto tall ships headed for destinations such as South Africa, the United Kingdom, and the eastern states of Australia (South West Development Commission 2014). Jarrah was frequently used for road paving blocks and fencing and karri timber was used for railway sleepers, boat building, and jetty construction.

Woodchips were first exported from the region in the 1970s to supply paper pulp manufacturers (South West Development Commission 2014), and it was around this time that community attitudes towards logging began to change. After the last steam-driven timber mills were closed in the late 1970s, the locations of large mills were limited to Deanmill (jarrah), Diamond Mill at Manjimup (woodchips), and Pemberton (karri). Long term forest campaigner Dr Beth Schultz recalls seeing a poster pasted up on a pole in Nedlands inviting people to attend a meeting about the government's proposal to begin woodchipping in the Southwest. She and a friend went along to the meeting at which the Campaign to Save Native Forests (CSNF) was born (Schultz 2015).

In *Contested Country: A History of the Northcliffe Area, Western Australia*, Patricia and Ian Crawford note that, in the early 1970s, growing public concern about the loss of Southwest forests and the manner in which they were being managed led to changes in the way land was articulated. The word "conservationist" began to replace "naturalist," and the word "environmentalist" began to be adopted. Terms like "green" and "wilderness" also crept into the vernacular while the words "conservation" and "environment" began to appear even in government documents (Crawford and Crawford 2003). Peter Robertson argues that it was as a result of the forest campaigns of the 1990s and beyond that the term "old-growth forest" has come into common parlance and is now used in forestry department documents in the region (Robertson 2015).

Forest Rescue: Historical Context of the Giblett Forest Campaign

There has been longstanding public concern over forestry management in Western Australia. Chapman argues that the forest protest movement in Western Australia is characterized by two periods of activity: the first in the 1890s with public concern expressed over the commercial destruction of forests, and the second from the 1950s when conservation groups formed to lobby governments and publicize their concerns over forestry practices (Chapman 2008). Elim Papadakis (1993) suggests that this historical environmentalism was a nationwide concern when he writes that the Australian environment movement began in 1795 with the efforts of Governor Hunter to save the trees along the Hawkesbury River. Earlier attempts by Governor Phillip to preserve a fifteen-meter (nine-mile) green belt of trees along the Tank Stream in Sydney Cove between 1789–90—when he forbade settlers to cut down trees, put up buildings, and keep or slaughter animals—predate Hunter's acts of conservation (Cathcart 2009). However, it could be argued that the Australian environment movement began with the original inhabitants of the many countries that were to become Australia: the Aboriginal peoples who practiced landcare for at least fifty-thousand years before Hunter and Phillip happened along.

In Western Australia, Ron Chapman argues, the WA forest protest movement has been characterized by an ability to adapt, evolve, and change its organizational structures and campaign strategies to meet evolving social and political climates. He identifies two distinct periods of forest protest in WA. The first period, beginning in 1895, he describes as "non-confrontational" and conciliatory in its bid to conserve native forests. The second period, beginning in the mid 1950s, evolved through several incarnations. From the 1950s to mid-1960s, the focus was on the attempt to preserve urban bushlands. Chapman describes

the 1970s as a transitional phase in which forest activists adopted a more asser-
tive mode of protest and, in the 1980s, forest conservation groups formed an
alliance with the Australian Labor Party (ALP) to collaborate on forest protection policy (Chapman 2008).

In early 1979, the WA Campaign to Save Native Forests (CSNF) and the
Southwest Forest Defence Foundation (SFDF) activists peacefully occupied
the site of the proposed expansion of bauxite mining in the jarrah forests at
the Wagerup near Waroona in the state's south-west corner. Former Greens
Member of the Legislative Council (MLC), Giz Watson, recalls being a part of
this group that set up camp on the refinery grounds and were arrested for tres-
passing. The group was trained in non-violent action by members of the Quaker
church and also based their training on the anti-nuclear protests happening
internationally such as Seabrook in the United States and Whyl in Germany,
as at the time as there were no precedents to this kind of peaceful occupation of protest sites in Australia (Elix 2011). Over the ensuing years, when it
appeared that conventional lobbying was failing to protect the environment,
this kind of direct action became increasingly important in campaigns such
as the Cockburn Women's Peace Camp, the Franklin River blockade, and Pine
Gap and Roxy Downs anti-uranium protests.

In the 1980s, woodchipping and clear-felling were in full swing in the
Southwest. Clear-felling was, and remains today, the preferred method of
logging because it is the cheapest way to produce large volumes of chip logs.
The forestry industry's own data shows that ninety percent of fauna in a logging coupe is killed during clear-felling.[3] By 1990, conservation groups' frustra-
tion with the way Southwest forests were being managed led to a big public
meeting in Bunbury, at which the WA Forest Alliance (WAFA) was formed.
WAFA comprised twenty different member groups[4] and worked on a number

---

3 Clear-felling is the removal of all trees from a designated forest harvest area known as a
   "coupe." After the trunk section of a tree that is suitable for sawmilling or woodchipping is
   removed from the coupe, all other forest residue such as branches, foliage, and bark, termed
   "slash," is left to dry out and then burned either by people on the ground or from helicopters,
   which drop ping pong balls injected with a napalm-type substance that self-ignites. The regeneration burns usually result in all organic matter in and around the coupe burning Forest
   Network (2001).

4 Augusta-Margaret River Friends of the Forest, Australian Conservation Foundation, Balingup
   Friends of the Forest, Blackwood Environment Society, Busselton Peace and Environment
   Group, Bridgetown-Greenbushes Friends of the Forest, Campaign to Save Native Forests,
   Crowea Committee, Coalition for Denmark's Environment, Conservation Council of WA,
   D'Entrecasteaux Defence Group, Dwellingup Greenbelt Committee, Friends of the Balckwood
   Valley, Great Walk Network, Leeuwin Conservation Group, South-West Environment Centre,

of levels to activate community support, attract media attention, and develop a focused non-violent direct action campaign.

Between 1994–95, about fifty hectares (one-hundred-and-twenty-three acres) of Giblett forest was cleared and woodchips made from the timber were exported. This action by the department of Conservation and Land Management (CALM) and woodchipping company Bunnings breached the conditions of Bunnings' woodchip license because of Giblett's high conservation status but no penalties were imposed. Karri/marri woodchips are considered to be the second lowest of eight grades of hardwood bought by Japanese pulpmills (Schultz 1994). The fact that WA's forests were being sold off for cheap paper products was seen by many in the community as environmental and economic vandalism. The Giblett forest rescue campaign began as a non-violent direct action (NVDA) in 1994 in response to what was seen in the forest conservation movement as gross mismanagement of WA's forest resources.

Chris Lee was one of the activists participating in the 1994 action, and recalls that there was a lot of resistance to undertaking direct action from within WAFA itself (see Figure 8.1). He believes that many of the established academics in WAFA did not have a conception of what NVDA was all about and were frightened by the idea of potentially breaking the law. He comments that "they were into doing the research and proving the case against woodchipping which of course is perfectly valid and part of the whole campaign but they were opposing NVDA which was what many of us believed was what really needed to happen" (Lee 2015). Disagreement in the group over direct action led to a compromise being reached, which was that, rather than going straight in and blockading the forest, WAFA would seek out some private land nearby and camp for a month to undertake NVDA training and do symbolic actions in the forest.

Lee comments that Giblett came out of an agreed strategy about the need to communicate concerns on forest management practices to the broader public. It was agreed that clear-felling, woodchipping, and the decline of old-growth native forests could not be talked about in the abstract, and there was a need to make these issues concrete for the community. It was finally agreed that the best way to do that was to adopt certain forest blocks that would be symbolic of the wider argument to end logging of old-growth forests in the Southwest. It was agreed to adopt Giblett, Hawke, and Jane blocks as a focus for action (Duggie 2015; Lee 2015; Schultz 2015).

South-West Forests Defence Foundation, Sustainable Agriculture Research Institute, The Wilderness Society, Warren Environment Group.

Local farmer Charlie Choderovsky, who lived adjacent to Giblett Block, offered the use of his land and a month-long forest camp was set up. Lee describes how the camp was established and run:

> Charlie said we could camp there so this camp emerged. At times there
> were a couple of hundred people camped in his paddock. The impor-
> tance of this camp was that it broke the impasse around direct action
> in WAFA. It enabled us to spend a month there doing stuff that was not
> going to offend anybody mostly and to reassure the old guard of WAFA
> that it was ok to do that. There was a group called Groundswell who were
> involved in NVDA and teaching those skills. There was a bunch of them
> who came down to do workshops and enable the whole discussion of
> non-violence. That was so important because from that month on, that
> was what the campaign was based on and it was absolutely rock solid.
> There were people there, like Via and Trevor Clarke, and Tom Bombadil.
> These people re-emerged when we went back in 1997. From day one when
> the platform went up in '97 the camp was absolutely based upon NVDA
> and there were enough of the people there who had been at the camp
> at '94 who came immediately back in '97 to help set that up. People in
> Groundswell, like Brenda Roy, Cheryl Lange, Sky, and Adrian Glamorgan,
> and people who had been involved in the Cockburn Women's Peace
> Camp [were there].
>
> LEE 2015

James Duggie was also at the 1994 camp and acted as a facilitator. In 1990,
Duggie had finished a physics degree and travelled overseas where he witnessed communities struggling to deal with pollution and poverty. He came
to the conclusion that Western Australia's economy and the world in general
were following an extremely unsustainable pathway. This motivated him to
want to be part of the solution rather than the problem. He became involved
in the WAFA forest campaign after becoming familiar with the issues around
management of Southwest forests. He recalls that he became convinced by
the argument put out by Southwest forest activists that management of
forests, and silviculture practices in general, were not sustainable in the
long run, either ecologically or from a silvicultural productivity perspective.
Duggie committed himself to try and change public opinion on forest management practices and comments that, "Giblett came out of the strategy behind how we thought we could have the most influence. Peter Robertson and
Beth Schultz had lot of input into strategy and the way they worked was quite
brilliant" (Duggie 2015).

Throughout the 1990s, WAFA, with its 25–30 member groups, convened regularly for strategy meetings and to collaborate on the Southwest forests campaign. WAFA was largely made up of people living in the region, and there was a lot of support from locals. Every two to three months, a meeting was held in various Southwest towns. A different member group of WAFA would host the meeting with representatives from the different groups, with up to forty people attending. Often the same core people would attend the full weekend meeting to update each other on everyone's individual campaigns, and to strategize about WAFA's broader approach. WAFA ran on consensus decision making, and gatherings also included training in consensus processes and NVDA. Member groups of WAFA, spread throughout the Southwest, would then return to their member bases, undertake their individual actions, and meet together every few months.

WAFA's key arguments were based on diminishing areas of native forest. In 1994, one-half of the pre-European area of karri forest and one-third of jarrah forest had been permanently cleared, mainly for agriculture, and only one percent of WA was covered by tall forest. Of that one percent, less than one-fifth was in conservation reserves. Western Australia had an estimated one-hundred-and-twenty-five thousand hectares (three-hundred-and-nine-thousand acres) of old growth karri and marri forest left, most of which was listed as available for clear-felling. Between one-thousand-and five-hundred to two-thousand hectares of karri and marri forest were being clear-felled in WA each year (Schultz 1994). Clear-felling kills up to ninety percent of the native birds and animals that live in a clear-fell area, as it destroys their habitat and food supply, and the destruction of ancient trees with hollows, which take up to three-hundred years to develop, puts at least fifty-one species at risk of extinction (Schultz 1994; Robertson 1997).

It was part of WAFA's campaign to try to debunk the idea that conservation of forests would destroy the timber industry. They asserted that, although more than fifty percent of all logs taken from clear-felled forests end up as woodchip, woodchipping only employed two-and-one-half percent of the forest-based work force. Furthermore, woodchipping competes with sawmills by taking logs they could use for value-added saw logs. Woodchipping also destroys the natural resources upon which industries such as tourism and bee-keeping depend (Schultz 1994).

Duggie and Lee remember WAFA meetings as being challenging and, at times, difficult, as the groups tried to reach consensus on strategies and actions (see Figure 8.2). Duggie comments that "there were so many ideas and strategies to discuss. The meetings would run for the full day on the Saturday, we would socialize on Saturday night, and then on Sunday morning we would have to finish and sort out agreement and consensus" (Duggie 2015).

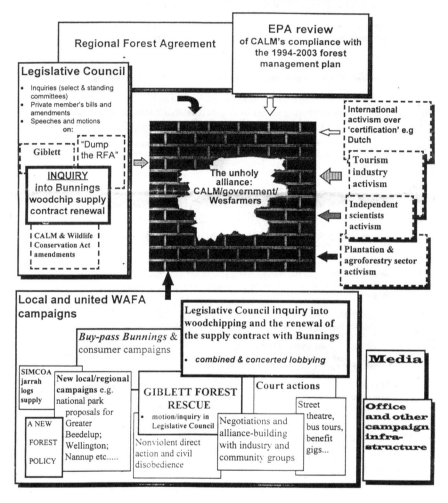

FIGURE 8.2 *"Schematic Guide to the WAFA Forest Campaign" (from Robertson 1997).*

In response to WAFA's activities, the Forest Protection Society (FPS) was formed. The FPS claimed to represent the views of the timber workers but its postal address was the same as the Forest Industry Federation of Western Australia (FIFWA). The FPS tried to create a sense that it was a grass-roots campaign consisting of on-the-ground timber workers, but Maddock and Duggie claim that most people understood that it was an industry-funded group, as it shared an office and was funded by FIFWA (Duggie 2015; Maddock 2015). During the Giblett blockade, the FPS set up a parallel tree sit in a patch of re-growth forest about twenty kilometers (twelve miles) away from Giblett. They built a luxury platform and spent a few nights in it to attract media coverage.

Not a Holiday: Giblett Forest Rescue 1997

In one radio interview I did, possibly with Howard Sattler, the interviewer asked 'isn't this just a whole bunch of dole bludgers going down to the Southwest for a holiday?' I replied, 'this is the middle of winter, it's been raining every day, it's cold, it's wet, it's uncomfortable, this is not a holiday! People are down here, they have taken leave from their jobs, come from all walks of life, to show their commitment for this forest; to advocate for better management of our forests, and to save this old-growth forest. It is not a holiday'.

DUGGIE 2015

In 1997, Dr Beth Schultz, chair of the Conservation Council of Western Australia (CCWA) and founding member of WAFA, received an anonymous call from a road contractor in the Southwest. The caller informed Dr Schultz that they had just a put a logging road through the most beautiful karri forest left in the Southwest and wanted to know, "what you greenies are going to do about it." Along with Peter Robertson from WAFA, Schultz drove down to the site of Giblett Block and found a fully made logging road through the old-growth karri forest, signaling that logging activities were again set to commence in Giblett (Schultz 2015). Except for fifty hectares (one-hundred-and-twenty-three acres) clear-felled in 1994, the forest was all old-growth with numerous very large ancient trees containing nesting hollows, which provide habitat for the fifty-one species of native birds and animals that require tree hollows for breeding (Schultz 1997).

At the same time, Pemberton local Roger Cheeseman was driving through Giblett on one of his regular trips through the area and saw that the department of Conservation and Land Management (CALM) had started scrub-rolling in the forest in preparation for clear-felling. He telephoned Chris Lee of the Margaret River Friends of the Forest. This was during the time before widespread use of emails, Twitter, and Facebook, so Chris Lee got on the phone to his friend Wade

Freeman, who was a climber, and asked if he could put a tree platform up in the forest. Wade agreed to help on the proviso that Chris would guarantee to stay up there for three weeks. Having never done anything like that before, Lee agreed and along with Wade Freeman started work on the platform the next day. The tip-off happened on a Tuesday, and, by the Friday of the same week, they had the platform and all the equipment that they needed stacked on top of a car and they were on their way to Giblett. Chris Lee (2015) remembers that:

> We met Wade on the Saturday, went in, and chose a tree and then on Sunday morning we hoisted up the tree platform watched by two of the local police who were standing there in the logging coupe watching. They didn't try to stop us. They did not know what to do. I guess they had no law to charge us with at that time. I don't think there was any law against tree climbing or anything like that. By Sunday afternoon I was up in the tree. I actually stayed there for five weeks and during that time there was this whole camp which grew up underneath the tree. By the end of the five weeks there was a queue of people who wanted to get up in the tree so I came down and someone else went up. That platform was occupied continuously for eight months. I went up in the first weekend in May '97 and we stayed there until early December.

This was the beginning of Western Australia's first long-term forest blockade, an event which arguably had a significant impact on both subsequent forest policy and on future activism in the state.

As a Southwest local, I visited Giblett several times in 1997 to offer food and financial support to the blockade. I also wrote an article about the camp for *Nova Journal* in which I described the scene as reminiscent of the set of the anti-war film *All Is Quiet on the Western Front* (1930). Tarpaulins, banners, and various makeshift shelters were strung up among the dead carcasses of huge fallen trees (see Figure 8.3). A column of smoke rose from the communal kitchen area, and people moved purposefully among the debris. It had the appearance of a desolate war zone (Chinna 1997). As a result of my support for the forest rescue, I personally experienced bullying and harassment in my own community of Quinninup, including people throwing stones at my house, as well as verbal abuse.

Non-Violent Direct Action

All of those interviewed for this chapter agree that the most important element in the success of Giblett and subsequent campaigns was the absolute

FIGURE 8.3   *"Camp at Giblett." 1997.*
        IMAGE COURTESY OF BETH SCHULTZ.

commitment of those involved to the principles of Non-Violent Direct Action (NVDA) (Duggie 2015; Lee 2015; Maddock 2015; Schultz 2015). One leading researcher in the field defines NVDA thus:

> Nonviolent action refers to those methods of protest, resistance, and intervention without physical violence in which the members of the nonviolent group do, or refuse to do, certain things. They may commit acts of omission—refuse to perform acts which they usually perform, are expected by custom to perform, or are required by law or regulation to perform; or acts of commission—perform acts which they usually do not perform, are not expected by custom to perform, or are forbidden by law or regulation from performing; or a combination of both.
>
> SHARP 1980

Duggie asserts that NVDA is about bearing witness to the commitment of people who believe that things needs to change, and their willingness to break the law and resist current institutional practices in their commitment to change (Duggie 2015). One example of NVDA tactics that was put into practice during the campaign was when American Buddhist teacher and activist Joanna Macy visited the region to work with forest activists and teach them the Elm Dance. The dance consists of four beats of movement, alternating with four beats of swaying in place. When swaying in place, participants imagine that they can feel the energy from the heart of the Earth spiraling up through the floor into their bodies. When the energy reaches the heart, people are instructed to send it out for the healing of the elms and all beings.

In her autobiography *Widening Circles*, Macy writes about her time in WA's Southwest forests:

> [...] 'the police don't arrest us while we're dancing' say our friends in Australia who have incorporated the Elm dance into their direct actions to protect the last stands of old growth forest [...] Under ancient karri trees, the tallest and most beautiful of the eucalypts, I have seen a dance encircle a bulldozer and bring it to a halt.
>
> MACY 2000

Leith Maddock recalls an incident after one of Macy's NVDA workshops in which activists had learned the Elm Dance. Macy had told the forest activists that they were like the Tibetan "Shambala Warriors" and that their weapons were compassion and insight. When being questioned by police about their actions, the young activists told the police that they were "Shambala Warriors and their weapons were compassion and insight." This was later reported in the local *Manjimup-Bridgetown Times* as "Shambala Warriors defend forests with weapons of compassion and insight" (Maddock 2015).

While the principles of NVDA directed all actions at Giblett, Chris Lee recalls that things frequently got heated between loggers and rescuers, and the camp facilitators had to talk things down. He comments that:

> there would be a roar in the night and a couple of ute loads full of drunken guys from the logging industry would arrive, and they would get out of their utes and come and piss in the fire and abuse us and throw things around. The young women at the camp like Jess Beckerling and Via and Emma Belfield, they would leap to their feet and run towards where the utes had stopped in order to get there first and to engage with the guys to practice what they completely believed in which was NVDA.
>
> LEE 2015

Maddock comments that here was an underbelly to the camp: "there was the Fluffies and the Hardcore, division in the camp, NVDA crew were Fluffies. The Hardcore worked hard, dug cars into the road, and filled them with cement. They did amazing things. But I seem to remember that the Fluffies did

FIGURE 8.4 *"Forest Gathering Poster."* 1997.
IMAGE COURTESY OF BETH SCHULTZ.

a lot more dishes than the Hardcore" (Maddock 2015). Maddock and Lee re-member that these divisions did not seem to be gendered and women were represented in both groups.

The animosity between timber workers and conservationists was compounded when, in 1999, an act of arson occurred in a Northcliffe craft shop owned and run by community members sympathetic to the conservation of old-growth forests. In September 1999, two men from Manjimup, Michael Grimes, 52, a house painter and Gavin Cain, 42, a timber worker, fire bombed the Northcliffe craft shop causing extensive damage to the heritage building. On the same day around forty men dressed in balaclavas and wielding baseball bats and crowbars threatened conservationists at Wattle Block protest camp, resulting in nine men being charged for violent and threatening behavior (Sinclair 1999).

However, the appeal of non-violence was not restricted to the forest rescuers but was embedded in more widespread community values. As a result of attacks by locals on the forest rescue camp at Wattle Block and other acts of violence, two local Manjimup women from different sides of the forest debate organized a gathering in Manjimup to promote zero tolerance for violence in the community. Shirley Curo whose husband worked in the timber industry commented that, "I could have wept when I heard about the business out at Wattle. No problems are solved with violence. It achieves absolutely nothing" (Sinclair 1999).

Campaign Momentum

> They are Warriors of the undergrowth, dreadlocked fairies at the bottom of God's garden. Call them what you will—ferals, hippies, greenies, dole bludgers—there's no doubting that the bedraggled sons and daughters of predominantly middle class WA have played a significant role in swinging public opinion against the destruction of old-growth forests.
> BARASS 1999

During Giblett forest rescue, campaigners were not sitting idle under a tarpaulin in the forest, but were instigating a range of campaign strategies and activities (Figure 8.4). Forest rescuers set up a "listening post" in the nearly town of Pemberton to promote discussion and dialogue in the community on forest issues. Rallies, walks, and fundraisers were held in Perth and many support groups shot up, including "Friends of Giblett Fremantle," "Doctors for Old Growth Forests," "Liberals for Forests," and "Suits for Forests." In addition, a series of coordinated actions took place in the Southwest, in which activists shut down woodchip mills, blockaded the woodchip train, and shut down the Bunbury Port.

Chris Lee recalls the extraordinary logistics involved in getting one-hundred people up to the Bunbury port before dawn in the days before mobile phones and the Internet, including one van load of young activists who did not have enough petrol to get to Bunbury. This group ended up being rescued by someone from CALM and being given fuel from the CALM depot! At Bunbury, Chris Lee recalled:

> We had people locked on or suspended from the woodchip ship in Bunbury Port, onto the conveyer belt that loaded the woodchips, and the woodchip train. All three of those elements came together in Bunbury. One-hundred people descended on the place at 2 am and locked on so all these three elements happened.
>
> LEE 2015

These kinds of actions between bodies and machinery were repeated at Giblett, Jane, Hester, Kingston, and Wattle blocks.

In addition to these risky embodied activities, the campaign attracted high-profile respectability with supporters such as West Coast Eagles (Australian Football League) coach Mick Malthouse, and fashion designer Liz Davenport. On Arbor Day in June 1997, former Eagles star Chris Turley visited Giblett and climbed the tree platform to show his support for old-growth forests. WAFA's campaign was successful at reaching right into the heart of conservative politics with even Dame Rachel Cleland whose husband Donald Cleland was one of the founding members of the Australian Liberal Party, engaging in NVDA action down in the forest. Many liberal and national parliamentarians supported the campaign and, in the 1990s, the Nationals broke away from the Liberals to support protection for old-growth forests (Robertson 2015). One action by the group "Suits for Forests" entailed anti-logging supporters who worked in the heart of Perth's central business district (CBD) all converging on St George's Terrace (the main thoroughfare through Perth) at a designated time to stand in the street and telephone the premier Richard Court, jamming his phone lines and making a powerful statement about widespread opposition to logging old-growth forests.

An effective strategy used to engage support for the forest campaign was to show people, first-hand, the effects of clear-felling on old-growth forests. Lawyer Peter Rattigan purchased a bus and began conducting "Real Forest Tours," taking participants into the heart of forestry activities. Leith Maddock became deeply involved in the Southwest forest campaigns of the 1990s after her first trip to the forest on one of Rattigan's bus tours. She recalls:

> I thought, 'oh these hippies are doing a great job camping down here', but I hated camping and couldn't wait to get back to Freo [Fremantle].

Then Peter very cunningly drove the bus through a burning clear-fell. There is nothing like a burning clear-fell, I mean, it's just horrendous [...] it was on fire and there was smoke. I clearly remember looking out the window of the bush and thinking 'No I don't think so'. That's when I started Friends of Giblett Fremantle. There was a whole crew in Freo of other people who had been on that bus, we used to meet at Gino's [restaurant] in the beginning and have coffee-fuelled meetings.

MADDOCK 2015

Friends of Giblett Fremantle (FOGF) was a major source of advocacy and support for the blockade in the south, fund raising and collecting donated goods such as tents, tarpaulins, ropes, food, and money. FOGF directly asked people for help, and activists needed some interesting objects, such as thumb cuffs, to lock their bodies onto forestry equipment. Maddock remembers driving around all the sex shops in Perth to ask if they would donate a few pairs of thumb cuffs each, and recalls that they were happy to donate them (Maddock 2015). Despite having a child, a job, and part-time study, Maddock used to go down to the forest on weekends. She comments that "I certainly did turn my life over to it I would say" (Maddock 2015). The impact of witnessing forestry "management" first-hand was a highly effective strategy for extending community support for the campaign.

Another major component in the plethora of WAFA actions was the By-Pass Bunning's campaign. By-Pass Bunnings was a hugely successful community campaign which called upon consumers to boycott the huge hardware chain Bunnings, which was a major shareholder in woodchipping companies and was responsible for ninety percent of WA's woodchip production (The Wilderness Society 1997). The Wilderness Society and WAFA coordinated this highly effective campaign, which resulted in Bunnings selling its interests in logging. Beth Schultz credits Cheryl Edwards, then-state environment minister in the Court government, as being vital to the success of the Giblett blockade in that she did not put a temporary control order on Giblett, meaning that activists would not be arrested simply for being in the forest.

Influence of Giblett: Now and the Future

At the end of the seven-month blockade, Giblett was never logged and is now part of the Greater Beedelup National Park. Visitors to the area today can listen to commentary on the local tourist radio station and read a story board with photographs, which describe how Giblett was saved from logging. However, the impact of the Western Australian forest campaigns of the 1990s, for which

FIGURE 8.5   *"Proposed Greater Beedelup National Park." 1996.*
IMAGE COURTESY OF WAFA.

Giblett was a pivotal event, have had a lasting impact on the politics of forest management and, indeed, upon other campaigns including anti-fracking, wetland conservation, and refugee advocacy.

James Duggie is now Principal Policy Officer of the Adaptation Climate Change Unit of the WA Department of Environment and Conservation and leads the team that develops adaptation policy, implements adaptation projects, and provides advice on climate change adaptation to Government and external stakeholders. Duggie comments that Giblett was part of a multi-year campaign, so it is difficult to point to one particular event and say that made all the difference. He suggests that all of the actions together built the momentum, resulting in enough awareness and pressure for things to change, but the blockades were important because they received widespread media coverage.

Leith Maddock and Chris Lee are now married and live in an intentional community in Denmark in Southwest WA. Chris and Leith have bought land together with some of the friends they made at Giblett and have planted twenty-thousand trees. Chris teaches young people to sail and Leith is working as a refugee advocate. Both Maddock and Lee maintain that Giblett enabled the development of great life skills in the participants and many of the young people who were at Giblett have gone on to be successful in their chosen fields:

> There are so many people who came down to the forest, 17/18 year olds dropping out of university or school or whatever and are now doing absolutely amazing things with their lives. Jess Beckerling was voted Young Western Australian of the year, and Ben Coin studied at Oxford and is now a human rights lawyer. If you made a map of the people who were at Giblett and where they all are now you would find that they were a bunch of high achievers. Most of the young crew we were working with were privileged, educated middle class kids and very motivated and driven. They have become high-achieving adults all working in advocacy and have stayed with their ideals.
>
> MADDOCK 2015

Most participants feel that the success of Giblett "rested on the shoulders of giants" (Lee 2015, Maddock 2015). The campaigns of the 1970s and '80s against the bauxite mining in jarrah forests and the anti-woodchip campaign of the 1970s were very powerful and put a huge amount of pressure on the government of the day. While they did not succeed in stopping the commencement of woodchipping, they raised extensive community and political awareness of forest issues. Peter Robertson himself was present at the historic Franklin River blockade in Tasmania in the 1980s, bringing skills and experiences honed there to the Giblett campaign. Robertson argues that, without those campaigns, the successful campaigns of the 1990s would not have been possible.

Peter Robertson notes that the Giblett blockade was a very significant part of the overall campaign to stop old-growth logging because it demonstrated the degree of commitment that campaigners had to see the campaign through to its conclusion (Robertson 2015). Robertson also reiterates there are still people such as Jess Beckerling out in the forests campaigning largely because of the empowerment they gained from Giblett. He comments that "once you have done something like that, you always know that something similar is possible again in the future. Giblett was an inspiration and an encouragement to act" (Robertson 2015).

Whether logging in old-growth forests has ceased in Western Australia is still a contested narrative. The forest industry continues to log Southwest forests, and there are still activists on the ground working to prevent it. According to Beth Schultz, around five-hundred hectares (one-thousand-two-hundred-and-thirty-five acres) of old-growth forests are still being logged in WA. Twenty percent of these are saw logs, and the rest goes to woodchips (Schultz 2015). The 2014–23 Forest Management Plan states that no old-growth forests have been, or are being, logged since the Regional Forest Agreement (RFA) of 2001 that strongly divided logging, conservation, activist, and Indigenous communities in the Southwest (Conservation Commission of Western Australia 2013). This is disputed by Schultz, Robertson, and Beckerling who assert that logging, which is permitted in so-called "Two Tier" forest, is targeting what is essentially old-growth forest. "Two Tier forest" is a general term used to describe forest of mixed age and structure, comprising mature trees intermixed with younger re-growth trees that have arisen from regeneration following previous harvests or other disturbances (sometimes also referred to as "mixed species, mixed age forest" or "mixed species, uneven-aged forest") (Conservation Commission of Western Australia 2013). CALM has now been split into two departments: the Forest Products Commission (FPC) and the Department of Parks and Wildlife (DPAW). Because the FPC defines old-growth forest as forest where "industrial disturbance is now not evident, or one stump per hectare" (Conservation Commission of Western Australia 2013), even if only two trees have been removed, the coupe is available for logging. Previous logging practices that used selective felling will have left evidence of cut trees within patches of old-growth forest. In this way, old-growth logging is still occurring in Southwest forests (Schultz 2015). In 2013–14, old-growth forest in Challar and Helms blocks was logged, and protest action ensued.

Peter Robertson is now the Wilderness Society WA State Manager and continues to work on campaigns to protect the environment including the recently successful James Price Point No Gas campaign. He comments that people are still identifying patches of old-growth forest in the Southwest that were not mapped in the 2001 RFA, and the department has admitted that some of these areas do fit the description of old-growth. At the time of our interview (February 17, 2015), there were three Southwest forest blockades in progress. The then-Barnett Liberal government had recently released its draft forest management plan for the next ten years and Peter Robertson commented that WAFA tried to get better outcomes for the forests by attempting to protect a few key areas under imminent threat. He says that the forests we now enjoy in the

Southwest are in existence mainly due to fifty years of forest activism and that, in fact, without these campaigns to save areas such as Giblett and the Shannon National Park, the Southwest would be unrecognizable, as there would hardly be a skerrick of old-growth forest left to admire (Robertson 2015).

In October 2014, WAFA released a statement expressing shock and dismay that the WA Government logging agency, the Forest Products Commission (FPC), submitted false and misleading information yet nevertheless gained an environmental tick of approval from the world's leading timber certifying body, the Forest Stewardship Council (FSC). WAFA stated that they had "shown the certifying body that the information provided by FPC is demonstrably false" and called upon the FSC to withdraw the certification in order to maintain its credibility in Australia (WA Forest Alliance 2014).

Jess Beckerling was a key activist at Giblett, and is still working to try to protect old-growth forests in the Southwest. She is now convener of WAFA, and writes that FPC's mapping of two-tiered high conservation value forests provided for this application was rife with errors that misled the certifying body and the public. She comments that:

> Woodchipping six-hundred-year-old karri trees and destroying habitat for endangered species is in no way sustainable, no matter how FPC tries to spin it. We have found ancient nesting hollow trees, critical for the survival of black cockatoos, destroyed while the forest was being logged. At least fifteen threatened fauna species including red-tailed black cockatoos, mainland quokkas, and chuditch lose vital habitat during karri logging operations. The FPC's logging practices are paving the way to extinction for these animals that exist nowhere else on earth.
>
> BECKERLING QTD. IN WA FOREST ALLIANCE 2014

Apart from logging activities, other key threats to the long term protection of Southwest forests include climate change, wildfires, and prescribed burn fire regimes, phytophthora dieback (*Phytophthora cinnamomi*), and feral flora and fauna (Schultz 2015). FPC's own data indicates that climate change is a critical and compelling risk to the forests. The most recent Southwest forest management plan for 2013–23, which projects future forest yields and harvest plans, states that in the Southwest, the impact of climate change has been most apparent in a substantial drying trend, with significant decreases in rainfall, streamflow, and groundwater levels recorded in the last forty years. Decreases in groundwater levels have been particularly evident in the northern jarrah forest, associated with a fifteen percent reduction in rainfall since the mid-1970s (Conservation Commission of Western Australia 2013).

Projections by the Indian Ocean Climate Initiative (2008) indicate that, based on the period 1960–1990, rainfall in the Southwest will have decreased by between two to twenty per cent by 2030, and by between five to sixty percent by 2070. Summer temperatures may increase by between 0.5 to 2.1 degrees Celsius (C) by 2030, and by between 1.0 to 6.5 degrees C by 2070, and winter temperatures may increase by between 0.5 to 2.0 degrees C by 2030, and by between 1.0 to 5.5 degrees C by 2070 (Bates 2008). With lower rainfall and increased summer temperatures comes increased risk of wildfires and hot burn bushfires. DPAW's strategy around this increased risk is to conduct more controlled burning to reduce fuel loads. Beth Schultz is currently researching the impact of controlled burning of Southwest forests by DPAW, and maintains that there is strong evidence to suggest that regular burning reduces biodiversity, with the "long un-burnt—at least ten-year cycles" being the optimum period to maintain species diversity. Schultz maintains that short periods between burns in the karri forest actually increase fuel loads because they increase understory growth.

Western Australian forests are globally unique and their importance was recognized when the Southwest Australia Ecoregion was selected as one of the thirty-five Global Priority Places in a scientific review conducted by the World Wildlife Fund in 2007 (Gole 2006). The area has also been recognized by Conservation International as one of only thirty-four global biodiversity hotspots, based on the high number of unique plants and the high level of threats to the region (Conservation International 2016). In addition, the Southwest Australia Ecoregion biome type is the most threatened of the planet's eight biomes because of extremely high levels of biodiversity and severe threatens to its integrity (World Wildlife Fund 2015). This region has the highest concentration of rare and endangered species in Australia and should be a primary focus for biodiversity conservation both nationally and internationally.

The Forest rescue campaigns of the 1990s were a compelling community movement that mobilized people from across the political spectrum to move "above and beyond" their everyday lives, to step up and take action to prevent wide scale clear-felling of old-growth native forests. For many of those involved, the experience was empowering and life changing. For those on the ground like Chris Lee and Jess Beckerling, the Giblett blockade was an exercise in faith and commitment to the principles of NVDA, in which they were physically and emotionally absorbed, leaving behind their study, jobs, and homes for as long as it took to achieve the goals of the campaign to end old-growth logging in Southwest forests. Beth Schultz comments that there was "an exhilaration about it [Giblett], people working together. Giblett inspired people to use individual initiative and the objective of saving Giblett and old-growth forests activated and enormous amount of people; it gave them a focus, an inspiration to do things above and beyond" (Schultz 2015). Chapman writes that it was ultimately the ability of the campaigners to convince Western Australians to take action to protect old-growth forests that led to the success of the campaigns (Chapman 2008). Recent anti-protest legislation proposed in WA, which would reverse the onus of proof and carries maximum penalties of two years jail or a twenty-four-thousand-dollar (AU) fine and cost recovery for any police response, places considerable uncertainty over future actions such as Giblett, James Price Point anti-gas blockade, and the Love Makes a Way refugee advocacy sit-ins. However, in the spirit of Giblett, current WAFA convenor Jess Beckerling remarks that the laws are excessive but will not stop people from defending the environment (Emerson 2015).

Postscript

Anti-protest legislation failed to get debated in parliament prior to the 2017 WA state election in which the Barnett government was defeated in a massive swing to the WA Labor Party.

Bibliography

Barass, Tony. 1999. "Dreadlock Democrats." *The West Australian*, February 17.

Bates, Bryson, Pandora Hope, Brian Ryan, Ian Smith, and Steve Charles. 2008. "Key Findings from the Indian Ocean Climate Initiative and Their Impact on Policy Development in Australia." *Climatic Change* 89(3): 339–54. doi: 10.1007/s10584-007-9390-9.

Bunbury, Bill. 1983. Something Unique, Something Majestic: The Karri Forest of the South-west. In *Strangers in a Chosen Landscape: Spoken Word as History*. Perth, WA: Australian Broadcasting Commission.

Cameron, James. 1981. *Ambition's Fire: The Agricultural Colonization of Pre-Convict Western Australia*. Perth, WA: University of Western Australia Press.

Cathcart, Michael. 2009. *The Water Dreamers: The Remarkable History of Our Dry Continent*. Melbourne: Text Publishing.

Chapman, Ron. 2008. "Fighting for the Forests: A History of the Western Australian Forest Protest Movement, 1895–2001." PhD diss., Murdoch University.

Chinna, Nandi. 1997. "Giblett Block: The Saga Continues." *Nova* (September).

Collard, Len. 2007. "Wangkiny Ngulluck Nyungar Nyittiny, Boodjar, Moort and Katitjin: Talking About Creation, Country, Family and Knowledge of the Nyungar of South-western Australia." In *Speaking from the Heart: Stories of Life, Family and Country*, edited by Sally Morgan, Tjalaminu Mia and Blaze Kwaymullina, 261–78. Fremantle, WA: Fremantle Press.

Collins, David. 1789. *An Account of the English Colony in New South Wales*. London: T. Cadell, Jun. and W. Davies (Successors to Mr. Cadell) in the Strand.

Conservation Commission of Western Australia. 2013. *Forest Management Plan 2014–2023*. Perth, WA: Government of Western Australia.

Conservation International. 2016. "Southwest Australia." Accessed April 12, 2017. http://www.cepf.net/resources/hotspots/Asia-Pacific/Pages/Southwest-Australia.aspx.

Crawford, Patricia, and Ian Crawford. 2003. *Contested Country: A History of the Northcliffe Area, Western Australia*. Nedlands, WA: University of Western Australia Press.

Department of Conservation and Land Management. 2000. *Forest Facts: Old Growth Forest*. Perth: Government of Western Australia.

Duggie, James. Interview with Nandi Chinna. Personal interview. Perth, February 1, 2015.

Elix, Jane. 2011. "Giz Watson." Accessed April 12, 2017. http://janeelix.wordpress.com/2011/05/09/giz-watson-coming-soon/, http://janeelix.wordpress.com/2011/05/09/giz-watson-coming-soon/.

Emerson, Daniel. 2015. "Protesters Face New Laws." *The West Australian*, March 6.

Forest Network. 2001. "What is Clearfelling?" Accessed April 12, 2017. http://www.forestnetwork.net/Docs/clearf.htm.

Hallam, Sylvia. 1975. *Fire and Hearth: A Study of Aboriginal Usage and European Usurpation in South-western Australia*. Canberra: Australian Institute of Aboriginal Studies.

Hamling, Bruce. 1979. "Maurice Coleman Davies, the Timberman." In *Westralian Portraits*, edited by Lyall, Hunt, 68–72. Nedlands, WA: University of Western Australia Press.

Heathcote, Ronald. 1976. "Early European Perceptions of the Australian Landscape: The First Hundred Years." In *Man and Landscape in Australia: Towards an Ecological Vision*, edited by George Seddon and Mari Davis, 40–44. Canberra: Australian Government Publications Service.

Lee, Chris. Interview with Nandi Chinna. Personal interview. Perth, February 1, 2015.

Macy, Joanna. 2000. *Widening Circles: A Memoir*. Gabriola Island, Canada: New Catalyst Books.

Maddock, Leith. Interview with Nandi Chinna. Personal interview. Perth, February 1, 2015.

Moore, Bryce. 1993. "Tourists, Scientists and Wilderness Enthusiasts: Early Conservationists of the South West." In *Portraits of the South West: Aborigines, Women and the Environment*, edited by Brian De Garis, 110–135. Nedlands, WA: University of Western Australia Press.

Moore, George Fletcher. 1884. *Diary of Ten Years Eventful Life of an Early Settler in Western Australia: And Also a Descriptive Vocabulary of the Language of the Aborigines*. London: M. Walbrook.

Nannup, Noel. 2003. "The Carers of Everything." Accessed April 12, 2017. http://www
.derbalnara.org.au/katitjin#3.

Papadakis, Elim. 1993. *Politics and Environment: The Australian Experience*. Sydney:
Allen and Unwin.

Powell, Joseph. 1976. *Environmental Mangement in Australia 1877–1914*. Oxford: Oxford
University Press.

Robertson, Peter. Interview with Nandi Chinna. Personal interview. Perth, February 1,
2015.

Robertson, Peter. 2013. "How we Stopped the James Price Point Gas Hub." *New
Matilda*, April 15. Accessed April 12, 2017. https://newmatilda.com/2013/04/15/
how-we-stopped-james-price-point-gas-hub/.

Robertson, Peter. 1997. *A Schematic Guide to the WAFA Forest Campaign*. Perth, WA:
Western Australian Forest Alliance.

Schultz, Beth. Interview with Nandi Chinna. Personal interview. Perth, February 1, 2015.

Schultz, Beth. 1997. *About Giblett Forest*. Perth: Conservation Council of Western
Australia.

Schultz, Beth. 1994. *Why Giblett Must Stand: Broadsheet*. Perth, WA: Western Australian
Forest Alliance.

Sclater, John. 2001. *Lost Your Block? The Origins of WA's Forest Block Names*. Safety Bay,
WA: John Sclater.

Sharp, Gene. 1980. *Social Power and Political Freedom*. Boston: Porter Sargent Publishers.

Sinclair, Fiona. 1999. "Women Unite in No Violence Stand." *Pemberton Northcliffe
Community New*s 102.

South West Development Commission. 2014. "Forestry." Accessed April 12, 2017. http://
www.swdc.wa.gov.au/industries/forestry.aspx.

South West Development Commission. 2012. *Manjimup SuperTown: Townsite Growth
Plan*. Perth: Government of Western Australia.

Stocker, Laura, Len Collard, and Angela Rooney. 2015. "Aboriginal World Views
and Colonisation: Implications for Coastal Sustainability." *Local Environment:
The International Journal of Justice and Sustainability* 21.7: 844–65. doi:
10.1080/13549839.2015.1036414.

Statham-Drew, Pamela. 2005. *James Stirling: Admiral and Founding Governor of Western
Australia*. Nedlands, WA: University of Western Australia Press.

The Wilderness Society. 1997. *Forest Update*. Perth, WA: The Wilderness Society.

WA Forest Alliance. 2014. "FSC Credibility on the Line After Certifying Karri
Clearfelling." Accessed April 12, 2017. http://waforestalliance.org/fsc-credibility-
line-after-certifying-karri-clearfelling.

World Wildlife Fund. 2015. "Southwest Australia Ecoregion." Accessed April 12, 2017.
http://www.wwf.org.au/our_work/saving_the_natural_world/australian_priority_
places/southwest_australia/southwest_australia_ecoregion/.

Index

Printed in the United States
By Bookmasters